PAT Past Paper
Worked Solutions

Published by *RAR Medical Services Limited*
www.uniadmissions.co.uk
info@uniadmissions.co.uk
Tel: 0208 068 0438

PAT Past Paper

Worked Solutions

Samuel Putra

Rohan Agarwal

UniAdmissions

The Basics..

2006 1

2007 1

2008 3

2009 3

2010 4

2011 5

2012 6

2013 8

2014 9

2015 10

2016 11

2017 12

2018 13

Your Free Book 14

Oxbridge Interview Course...14

About the Authors

Samuel is currently a third-year DPhil candidate in Engineering Science at Trinity College, University of Oxford. He obtained his MEng degree from Oxford with First-Class Honours, graduating in top 4% of his class. Since his undergraduate years, Samuel has assisted many A level students with their university admissions, providing tuition for entrance exam preparation and guidance for interviews.

During his postgraduate study, Samuel holds a teaching role as Graduate Teaching Assistant at Oxford, providing tutorials for undergraduates in Chemical Engineering Modules. His research is focused on Sustainable Wastewater Treatment and Energy Recovery which has secured several awards including Best Poster in Water Category at Concawe Symposium 2017. In his spare time, Samuel enjoys going to the gym and playing football.

Rohan is the **Director of Operations** at *UniAdmissions* and is responsible for its technical and commercial arms. He graduated from Gonville and Caius College, Cambridge and is a fully qualified doctor. Over the last five years, he has tutored hundreds of successful Oxbridge and Medical applicants. He has also authored ten books on admissions tests and interviews.

Rohan has taught physiology to undergraduates and interviewed medical school applicants for Cambridge. He has published research on bone physiology and writes education articles for the Independent and Huffington Post. In his spare time, Rohan enjoys playing the piano and table tennis.

THE BASICS

What are PAT Past Papers?

Hundreds of students take the PAT exam each year. These exam papers are then released online to help future students prepare for the exam.

Where can I get PAT Past Papers?

This book does not include PAT past paper questions because it would be over 1,000 pages long if it did! However, all PAT past papers since 2006 available for free from the official PAT website. To save you the hassle of downloading lots of files, we've put them all into one easy-to-access folder for you at **www.uniadmissions.co.uk/PAT-past-papers**.

How should I use PAT Past Papers?

PAT Past papers are one the best ways to prepare for the PAT. Careful use of them can dramatically boost your scores in a short period of time. The way you use them will depend on your learning style and how much time you have until the exam date but generally four to six weeks of focussed preparation is usually sufficient for most students.

How should I prepare for the PAT?

Although this is a cliché, the best way to prepare for the exam is to start early – ideally by September at the latest. If you're organised, you can follow the schema below:

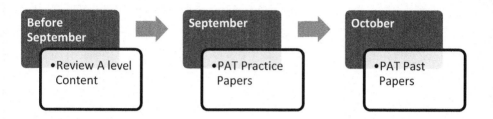

This paradigm allows you to minimise gaps in your knowledge before you start practicing with PAT style questions in a textbook. In general, aim to get a textbook that has lots of practice questions e.g. *PAT Practice Papers* (**www.uniadmissions.co.uk/PAT-book**) – this allows you to rapidly identify any weaknesses that you might have e.g. Newtonian mechanics, simultaneous equations etc.

You are strongly advised to get a copy of *'PAT Practice Papers'* which has hundreds of practice questions– you can get a free copy by following the instructions at the back of this book.

Finally, it's then time to move onto past papers. The number of PAT papers you can do will depend on the time you have available but you should try to do at least each paper once.

How should I use this book?

This book is designed to accelerate your learning from PAT past papers. Avoid the urge to have this book open alongside a past paper you're seeing for the first time. The PAT is difficult because of the intense time pressure it puts you under – the best way of replicating this is by doing past papers under strict exam conditions (no half measures!). Don't start out by doing past papers (see previous page) as this 'wastes' papers.

Once you've finished, take a break and then mark your answers. Then, review the questions that you got wrong followed by ones which you found tough/spent too much time on. This is the best way to learn and with practice, you should find yourself steadily improving. You should keep a track of your scores on the next page so you can track your progress.

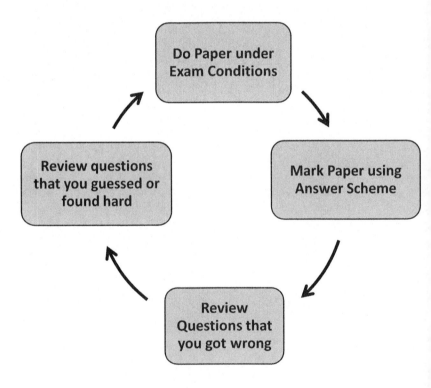

Scoring Tables

Use these to keep a record of your scores – you can then easily see which paper you should attempt next (always the one with the lowest score).

PAPER	1st Attempt	2nd Attempt	3rd Attempt
2006			
2007			
2008			
2009			
2010			
2011			
2012			
2013			
2014			
2015			
2016			
2017			

2006
Physics

Section A

Question 1: C
The term u here is initial velocity (m/s), whereas the term t is time in s. Hence ut is velocity x time which is displacement (m).

Question 2: A
First, we need to find the mass of the block. Mass = density x volume, whereas the volume would be the product of the three sides. Volume = 3 cm x 4 cm x 5 cm = 60 cm^3 = 6 x 10^{-5} m^3. The mass is then = 8570 kg/m^3 x 6 x 10^{-5} m^3 = 0.5142 kg. By definition, pressure is force/area; force is the weight of the block in this case (assume that gravitational acceleration is 10 $\frac{N}{kg}$). To get maximum pressure, the area needs to be minimised. Minimum area would be the product of the two smallest sides; 3 cm x 4 cm = 12 cm^2 = 12 x 10^{-4} m^2. The pressure is then (0.5142 kg x 10 $\frac{N}{kg}$) / (12 x 10^{-4} m^2) = 4285 Pa $\approx$ 4.3 kPa

Question 3: A
Assuming total energy is conserved, and it consists of only potential and kinetic energy, then at D the potential energy is at highest compared to the other points as it is at the furthest distance; hence kinetic energy here is at its lowest (which means that velocity is at its slowest).

Question 4: C
According to Archimedes' principle, the buoyancy force exerted on the object equals to the weight of the fluid displaced by the object. This force is then equal to the weight of the object as the object is in equilibrium (floating). Hence $mg = V_d \rho_w g$ where V_d is the displaced volume. Thus, $V_d = \frac{m}{\rho_w}$. If a mass (spoon of water) is added onto the boat; ie. $\rho_w \delta V$, then $V'_d = \frac{m + \rho_w \delta V}{\rho_w} = V_d + \delta V$. This means that more water is being displaced from the bath, increasing its water level. However, as the additional water was taken from the bath itself, then $V''_d = V_d + \delta V - \delta V = V_d$. So, the water level stays the same.

Question 5: B

In this question, the kinetic energy gained by the electron is derived from the electrical energy due to the applied potential. Electrical energy is charge e (assumed to be in coulomb) multiplied with potential V (J/c); ie. eV. Kinetic energy is $\frac{1}{2}mv^2$ which can be re-expressed as $\frac{p^2}{2m}$ (as $p = m \times v$). Hence $V = \frac{p^2}{2m} = \frac{h^2}{2m\lambda^2}$; rearranging gives us $\lambda = \frac{h}{\sqrt{2meV}}$

Question 6: D

For steady acceleration, distance travelled can be expressed as $d = \frac{(u+v)\,t}{2}$; where t is time, u and v are initial and final velocity accordingly. For the first car, $d = \frac{(0+20)\,t}{2} = 10\,t$; so $t = 0.1\,d$. Another car takes $t' = 2t$ to get to the same final velocity, hence the distance travelled by the second car $d' = \frac{(u+v)\,t'}{2} = \frac{(0+20)\,0.2d}{2} = 2d$

Question 7: C

The distances D are distributed as 1, 4, 9, 16, 25; the year lengths Y are 1, 8, 27, 64, 125, or can be written as $1\sqrt{1}$, $4\sqrt{4}$, $9\sqrt{9}$, $16\sqrt{16}$, $25\sqrt{25}$. Hence, we could see here that $Y \propto D\sqrt{D}$.

Question 8: C

The reading on the scale will show the Martian's weight divided by the gravitational strength at the surface of Venus, as the scale was designed to be used on Venus. Hence, reading $= R = \frac{m \times g_{mars}}{g_{venus}}$; hence $m = \frac{R \times g_{venus}}{g_{mars}} = \frac{93 \times 8.8}{3.8} = 215\,kg$

Question 9: B

First of all, mass will remain constant as there is no loss or gain of matters. Secondly, as the object shrinks, meaning that the volume is reduced; hence the density is increased.

Question 10: C

As it is a free fall, the velocity of the bag will increase from initial velocity 11 m/s due to earth's gravitational acceleration (10 m/s²). After 7s, the final velocity v will be $u + at = 11 + 10\times7 = 81$ m/s. Distance is then calculated using the formula $d = \frac{(u+v)\,t}{2} = \frac{(11+81)\,7}{2} = 322\,m$

END OF SECTION

Section B

Question 11

The important thing to note here is that brightness depends on the power delivered to the lamp; ie. $P = V^2/R$. Hence:

Lightbulb **a** will be dimmer than normal, as the potential from the cell is now divided into two identical bulbs.

Lightbulb **b** will be normal, as now we have two identical cells to power two identical bulbs.

Lightbulb **c** will be normal, as lightbulb **d** will be practically open-circuit (no current flow thus the lamp is off) because the potentials of the two cells act on opposite direction on **d**. Hence, this circuit virtually works the same way as the previous circuit.

Lightbulb **e** will be dimmer as the voltage is halved. Lightbulb **f** will be normal as the circuit is in parallel, hence the voltage across the lightbulb is the same as the cell's potential.

Lightbulb **g** will be brighter as the voltage is doubled from the two cells; lightbulb **h** will be normal as the left-hand side cell has no effect on it (only the right one contributes to its voltage).

Question 12

In this question, we have five unknowns (hence need five equations) which are the sides of red, green and blue cubes (r, g and b respectively), their density (ρ_{cube}) and the liquid density (ρ_{liquid}). We know that all cubes have the same density as all three of them float with half volume exposed. Furthermore, from this fact we also know that the liquid density has to be 2x the cubes' density; ie. $\rho_{liquid} = 2\rho_{cube}$ (1). This is in fact the first equation of the problem. The other four equations to solve for all variables are:

Statement (a):	$r + g = 35$	(2);
Statement (b):	$2g + b = 70$	(3);
Statement (c):	$r + b = 2g$	(4);
Statement (d):	$\rho_{cube} (r^3 + g^3 + b^3) = 20,000$ (in grams)	(5);

(3) minus (4) leads to $4g - r = 70$; then add with (2) gives us $5g = 105$; hence $g = 21$ cm $\rightarrow$ $r = 14$ cm; and $b = 28$ cm.

Substituting these values to (5) will then give $\rho_{cube} = 0.589$ g/cm³. The density of the liquid is then $2 \times 0.589 = 1.178$ g/cm³

Question 13

(a) For this part, it is obvious. The rocket reached its maximum velocity and started to slow down.

(b) Remember that acceleration is the derivative of velocity, and in this graph (velocity vs. time) it is the tangent of the curve. The bigger the gradient of the tangent, the higher the acceleration. It is then approximately just before point X (the tangent could be estimated as the straight line between point X and the point before it). Physically speaking, this is because at that point, all fuel has been used up so mass has decreased to its lowest value, hence acceleration is maximised.

(c) After point X, the velocity decreases steadily. Hence, the rocket decelerates uniformly.

(d) For velocity vs. time graph, remember that distance is obtained by integrating the curve. As the rocket is travelling upwards, the distance travelled is equivalent to the height reached by the rocket. Maximum height is obtained when the rocket stops moving, ie. its velocity reaches zero. Hence the straight line of the curve needs to be extended until it reaches the x-axis.

END OF SECTION

Section C

Question 14

(a) Electrical energy is obtained by multiplying power with time, ie. *Pt*. The relation is then:

$$Pt = Kinetic\ Energy\ (KE) = \frac{1}{2}mv^2; \quad v = \sqrt{\frac{2Pt}{m}} = \sqrt{\frac{2P}{m}}\,t^{\frac{1}{2}}$$

(b) Acceleration is the derivative of velocity *v* with respect to time *t;* whereas distance travelled is the integral of velocity over time. Hence:

$$a = \frac{dv}{dt} = \frac{1}{2}\sqrt{\frac{2P}{m}}\,t^{-\frac{1}{2}} = \sqrt{\frac{P}{2mt}}$$

$$d = \int_0^t v\,dt = \sqrt{\frac{2P}{m}}\int_0^t t^{\frac{1}{2}}\,dt = \sqrt{\frac{2P}{m}}\,[\frac{2}{3}t^{\frac{3}{2}}]_0^t = \frac{2}{3}\sqrt{\frac{2Pt^3}{m}}$$

(c) From the velocity expression we obtain in part (a), as t $\rightarrow$ ∞, v goes to infinity as well. This is not reasonable as infinite velocity is impossible.

(d) For acceleration *a*, as t goes to infinity, acceleration goes to zero. This is reasonable since acceleration cannot go on forever. As t is very small (close to zero), *a* goes to infinity which is also reasonable, showing rapid initial acceleration.

(e) For this case, *Pt* equals potential energy *mgh;* hence $h = \frac{Pt}{mg}$. Velocity of vertical movement can be written as $v_v = \frac{h}{t} = \frac{P}{mg}$

(f) For vertical movement, we have derived that $v_v = \frac{P}{mg}$. Hence, kinetic energy is $KE = \frac{1}{2}mv_v^2 = \frac{1}{2}\frac{P^2}{mg^2}$

Then, the ratio of kinetic energy to potential energy is $\frac{KE}{PE} = \frac{1}{2}\frac{P^2}{mg^2} x \frac{1}{Pt} = \frac{P}{2mg^2t}$; (remember that potential energy equals electrical energy).

Thus, kinetic energy can be ignored when *P* is low, and *t* is long.

END OF SECTION

Mathematics for Physics

Question 1

i) For this question, we need to realise that $x^2 - y^2$ is equal to $(x - y)(x + y)$. So, $2007^2 - 2006^2 = (2007 - 2006)(2007 + 2006) = 1(4013) = \mathbf{4013}$

ii) For this one, $1.001^6 - 1.001^5 = (1 + 0.001)^6 - (1 + 0.001)^5$. Using *Pascal's Triangle and Binomial Theorem*, you can expand the first term to $(1^6 + (6)1^5(0.001) + (15)1^4(0.001)^2 + \dots)$ and second term to $(1^5 + (5)1^4(0.001) + (10)1^3(0.001)^2 + \dots)$. However, only the first two terms of each expansion would be significant since the others are too small. Hence it is $(1 + 0.006) - (1 + 0.005) = \mathbf{0.001}$

Question 2

Gradient of a straight line between two points can be evaluated using the formula $m = (y_2 - y_1)/(x_2 - x_1)$ where (x_1,y_1) and (x_2,y_2) are the coordinates of the two points. In this case, the gradient is $(-2 - 8)/(5 - 4) = \mathbf{-10}$

Question 3

i) This one is quite straightforward – we just need to realise that $log_a (a^n) = n$. Hence, $log_e (e^{3x}) = 3x = 6; x = 2$

ii) In log, it can be written that $log_a b = c$ when $a^c = b$. Hence, $log_3 x^2 = 2$ can be written as $3^2 = x^2 = 9$.
So $x = \pm 3$

Question 4

To solve this question, we need to picture trigonometric functions (*sin, cos, tan*) in a right-angle triangle. For right-angle triangle as shown in figure below, *sin cos* and *tan* can be defined as $sin = \dfrac{opp}{hyp}, cos = \dfrac{adj}{hyp}$ and $tan = \dfrac{opp}{adj}$

Hence, $tan^{-1} (\frac{12}{5})$ can be thought as an angle θ of a right-angle triangle with $opp = 12$ and $adj = 5$. Using Pythagoras, we can work out the hypotenuse which is $hyp^2 = adj^2 + opp^2$; $hyp = \sqrt{5^2 + 12^2} = 13$.

Hence, $13 \, sin(tan^{-1} (\frac{12}{5})) = 13 sin \theta = 13 \, x \dfrac{opp}{hyp} = 13 \, x \, \frac{12}{13} = 12$

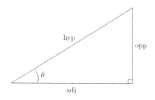

Question 5

i)   It is useful to remember how this curve looks like, which is as below

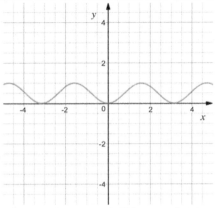

ii)   For this function, when $x = 0 \rightarrow y = -1$. When $y = 0$, x has no solutions. As $x \rightarrow \infty$ and $x \rightarrow -\infty$, $y \rightarrow 0$ (+ve) hence the *x-axis* is a horizontal asymptote. When $x = \pm 1$, y goes to infinity hence these gives two vertical asymptotes. The function then looks as follows:

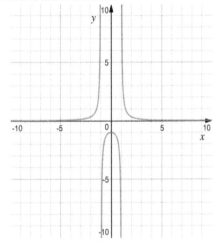

Question 6

In mathematical equations, the two statements can be written as follow:

$r_A = r_B + 1$ (1)

$\pi r_A^2 = \pi r_B^2 + 2\pi$ (2)

Substituting *(1)* into *(2)*, we get $r_B^2 + 2r_B + 1 = r_B^2 + 2$; $r_B = 0.5$ *cm*. Then, $r_A = $ **1.5 cm.**

Question 7

i)   The probability of each dice gives a six is $\frac{1}{6}$. Hence, the probability of all three dices give a six is $\frac{1}{6} * \frac{1}{6} * \frac{1}{6} = \frac{1}{216}$

ii)   In section (i) the probability of all three dices to give a six is $\frac{1}{216}$. Now, for the three dices to give the same number, the scenario could be 1,1,1 , 2,2,2, 3,3,3, and so on (6 scenarios in total), with each scenario having the same probability. Hence, the total probability is $6 * \frac{1}{216} = \frac{1}{36}$

iii) In this case, we do not want the first and the second dice to give a six, ie. anything but 6 hence the probability is $\frac{5}{6}$. Hence, the overall probability is $\frac{5}{6} * \frac{5}{6} * \frac{1}{6} = \frac{25}{216}$

Question 8

We know that $\frac{dV}{dt} = 1 \ cm^3 s^{-1}$. To find rate of growth of radius, ie. $\frac{dr}{dt}$, first we need to express V as a function of radius $\rightarrow V = \frac{4}{3}\pi r^3$. Differentiate this to give $\frac{dV}{dr} = 4\pi r^2$. Note that $4\pi r^2$ is the expression for sphere's surface area, hence $\frac{dV}{dr} = 4\pi r^2 = S = 100cm^2$. Finally, $\frac{dr}{dt} = \frac{dr}{dV} x \frac{dV}{dt} = \frac{1}{100} x 1 = \mathbf{0.01 \ cm \ s^{-1}}$

Question 9

The graph with all the lines are sketched below.

The total area can be split into two categories: (1) the rectangle between the x-axis and the $y = -2$ line; (2) the area between the curve and the x-axis. Area (1) is easy to calculate $\rightarrow$ *width x height* = 4 x 2 = 8.

To calculate area (2), we need to integrate the curve $\rightarrow \int_{-2}^{2} |x^n| \ dx = 2\int_{0}^{2} x^n \ dx = 2(\frac{1}{n+1}x^{n+1}|_0^2) = 2(\frac{1}{n+1}2^{n+1})$

So, the total area would be $A = 8 + 2(\frac{2^{n+1}}{n+1})$

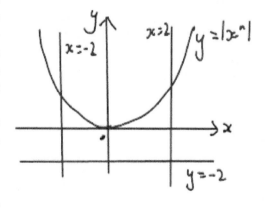

Question 10

i) $1 + e^y + e^{2y} + e^{3y} + \cdots$ is a geometric series with $r = e^y$. Since $r \ll 1$, then $S_\infty = \frac{U_1}{1-r} = \frac{1}{1-e^y}$

ii) You need to realise that $log_2 1 + log_2 2 + log_2 4 + \cdots + log_2 2^n$ is nothing more than $0 + 1 + 2 + \ldots + n$. Hence, this is an arithmetic series with common difference 1, first term 0 and total no. of terms $n+1$ (or first term = 1 and total terms n) $\rightarrow S_n = \frac{n}{2}(U_1 + U_n) = \frac{n}{2}(1 + n)$

Question 11

To identify stationary points, we need to differentiate the function and equate it to zero $\rightarrow \frac{dy}{dx} = 24 - 18x - 6x^2 = 0$. We can simplify this quadratic equation to $x^2 + 3x - 4 = 0$. Solve this quadratic equation by factorisation $\rightarrow (x + 4)(x - 1) = 0; x = -4 \; or \; x = 1$. To classify the type of stationary point, we need to differentiate it again $\rightarrow \frac{d^2y}{dx^2} = -18 - 12x$

At $x = -4 \rightarrow y = 5 + 24(-4) - 9(-4)^2 - 2(-4)^3 = -107; \frac{d^2y}{dx^2} = -18 - 12(-4) = 30$ which is $> 0 \rightarrow$ **(-4, -107) is a minimum point**

At $x = 1 \rightarrow y = 5 + 24(1) - 9(1)^2 - 2(1)^3 = 18; \frac{d^2y}{dx^2} = -18 - 12(1) = -30$ which is $< 0 \rightarrow$ **(1, 18) is a maximum point**

Question 12

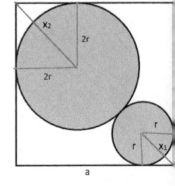

For smaller circle $\rightarrow x_1 = \sqrt{r^2 + r^2} = \sqrt{2}r$; for bigger circle $\rightarrow x_2 = \sqrt{(2r)^2 + (2r)^2} = 2\sqrt{2}r$
The total diagonal of the square $= x_2 + x_1 + 2r + r = 2\sqrt{2}r + \sqrt{2}r + 2r + r = \sqrt{2}a \quad \rightarrow \quad a = \frac{6+3\sqrt{2}}{2}r$

Total area of the square $= a^2$; area of both circles $= \pi(2r)^2 + \pi(r)^2 = 5\pi r^2$

So, ratio of both circles to square area $= \frac{5\pi r^2}{a^2} = \frac{5\pi r^2}{(\frac{6+3\sqrt{2}}{2}r)^2} = \frac{5\pi}{\frac{36+18+36\sqrt{2}}{4}} = \frac{20\pi}{54+36\sqrt{2}} = \frac{10\pi}{27+18\sqrt{2}}$

END OF PAPER

2007
Physics

Section A

Question 1: C

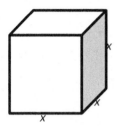

For a cube of side length x, resistance can be expressed as $R = \frac{L}{A} = \frac{x}{x^2} = \frac{1}{x}$. Hence, the resistance is inversely proportional to x.

Question 2: D
Astronauts experience weightlessness in ISS because they are accelerating at the same rate as the space station, hence they are orbiting the earth altogether and practically experiencing free-fall (hence weightlessness) but never reaching earth due to their forward velocity cancelling out the effect of falling down from earth's gravitational pull.

Question 3: A
$I = \frac{V}{R} = \frac{9}{100} = 0.09\ A = 0.09\ Cs^{-1}$. Given that charge of 1 electron is 1.6 x 10^-19 C, hence the no. of electrons per second is $= \frac{0.09}{1.6\ x\ 10^{-19}} =$ $5.6\ x\ 10^{17}\ electrons\ per\ second$

Question 4: C
For acceleration, as the slide gets less steep down the slide, his acceleration will decrease because gravity plays smaller part as he slides down. His speed, though, will keep getting faster because he still has acceleration.

Question 5: A
You need to remember that when springs are connected in series, their k value decreases. When two identical springs connected in series, each with spring constant k, the overall springs constant now become $k/2$.

Question 6: A
Power is calculated as $P = \frac{V^2}{R}$, so the power dissipated on the resistors are $\frac{V^2}{R_1}$ and $\frac{V^2}{R_2}$ respectively. Hence, the total power dissipated is $P_{tot} = \frac{V^2}{R_1} + \frac{V^2}{R_2} = V^2(\frac{1}{R_1} + \frac{1}{R_2})$

Question 7: B
As half of the drug is excreted by the body every two hours, and the isotope itself decays with half-life of two hours, then in overall it will take ½ *2 = 1 hour for it to be halved inside the body

Question 8: C
$$P = \frac{F}{A} = \frac{mg}{\pi r^2} = \frac{0.125x10}{\pi(0.001)^2} = 397,887 \, Pa \approx 400 \, kPa$$

Question 9: A
$\Sigma F = ma$; in this case, let's analyse the forces acting on the trailer. They are 1) the force on the towbar T pulling the trailer forward; 2) the friction of 2500 N on the trailer pulling it backward. Since both car and the trailer is accelerating at 4 m/s², then:
$$T - 2500 = (1000)(4); T = 6500 \, N$$

END OF SECTION

Section B

Question 10

a) If there is no wind/resistance, then time taken $t = \frac{d}{v} = \frac{300}{170} = 1.76\ hr \approx$
1 *hour* 46 *minutes*

Estimated arrival time is then 10:46 AM.

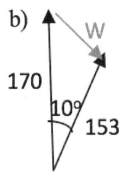

b) Using cosine rule; $W^2 = (170^2) + (153^2) -$
$2(170)(153)cos\ 10^o\ = 1079.3; W = 32.9\ km/h$

Question 11

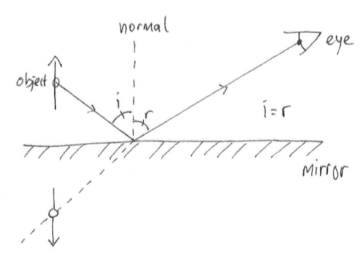

Question 12

We have three unknowns to solve, which are the size of the three caterpillars (mauve (m), lavender (l) and violet (v)).

From the statements in the question, we can deduce the following equations:

$$5m = 7v \qquad\qquad (1)$$
$$3l + 1m = 8v \qquad\qquad (2)$$
$$5l + 5m + 2v = 1 \text{ (in m)} \qquad\qquad (3)$$

From these three equations, we can find that $v = 0.05$ m. To travel this distance, it takes the lavender caterpillar 10 seconds; ie. $v_l = \frac{0.05}{10} = 0.005 \ m/s$. Then, we know that $v_m = 2v_l = 2(0.005) = 0.01 \ m/s$.

The total distance needs to be crawled by the mauve caterpillar is the circumference of the planet; $d = 2\pi r = 2\pi(1180000) = 7.41 \times 10^6 \ m$. Then, the total time needed is $t = \frac{d}{v} = \frac{7.41 \times 10^6}{0.01} = \mathbf{7.41x10^8 \ s}$

Question 13

When there is a current of 0.4 A flowing through the circuit, the voltage drop across the non-linear resistor can be calculated as $V_{nl}^3 = \frac{I}{0.05} = \frac{0.4}{0.05} = 8; V_{nl} = 2V$. The voltage drop across the fixed resistor then $V_{fix} = 9 - V_{nl} = 9 - 2 = 7V$. The resistance is then $R_{fix} = \frac{V_{fix}}{I} = \frac{7}{0.4} = \mathbf{17.5 \ \Omega}$

END OF SECTION

Section C

Question 14

(a) For an ideal pendulum, we can ignore the mass of the rod and treat the bob as a unit mass. Hence, L_{CM} in this case equals to L, and $I = ML^2 \rightarrow P = 2\pi\sqrt{\dfrac{ML^2}{gML}} = 2\pi\sqrt{\dfrac{L}{g}}$

(b) For a pendulum made of uniform rod, we know that the centre of mass would be the middle of the rod, ie. $L_{CM} = 0.5L$ and $I = \dfrac{1}{3}ML^2 \rightarrow P = 2\pi\sqrt{\dfrac{\frac{1}{3}ML^2}{gM\frac{1}{2}L}} = 2\pi\sqrt{\dfrac{2L}{3g}}$

(c) To find the effective length, we need to use mass-averaged effective length; ie. $L_{CM} = \dfrac{LM_b + \frac{L}{2}M_r}{M_b + M_r}$. I_{tot} would be the sum of the two I's, ie. $I_{tot} = M_bL^2 + \dfrac{1}{3}M_rL^2$

$\rightarrow P = 2\pi\sqrt{\dfrac{M_bL^2 + \frac{1}{3}M_rL^2}{g(M_b+M_r)\frac{LM_b+\frac{L}{2}M_r}{M_b+M_r}}} = 2\pi\sqrt{\dfrac{L(M_b+\frac{M_r}{3})}{g(M_b+\frac{M_r}{2})}} = 2\pi\sqrt{\dfrac{2L(3M_b+M_r)}{3g(2M_b+M_r)}}$

For an ideal pendulum, $M_r = 0 \rightarrow P = 2\pi\sqrt{\dfrac{2L(3M_b)}{3g(2M_b)}} = 2\pi\sqrt{\dfrac{L}{g}}$

For a rod pendulum, $M_b = 0 \rightarrow P = 2\pi\sqrt{\dfrac{2L(M_r)}{3g(M_r)}} = 2\pi\sqrt{\dfrac{2L}{3g}}$

(d)   We can assume it is an ideal pendulum, hence $P = 2\pi\sqrt{\dfrac{L}{g}}$; when there is an increase in temperature $\rightarrow P' = 2\pi\sqrt{\dfrac{L(1+\alpha\delta T)}{g}}$;

Change in period $\Delta P = P' - P = 2\pi\sqrt{\dfrac{L(1+\alpha\delta T)}{g}} - 2\pi\sqrt{\dfrac{L}{g}} = 2\pi\sqrt{\dfrac{L}{g}}(\sqrt{1+\alpha\delta T} - 1) = P(\sqrt{1+\alpha\delta T} - 1)$

For an accuracy of 1 second in 24hr $\rightarrow \dfrac{\Delta P}{P} = \dfrac{1}{24 \times 3600} = (\sqrt{1+\alpha\delta T} - 1)$

Given that $\alpha = 19 \times 10^{-6}$ K^{-1}, rearranging the above equation gives $\delta T = \mathbf{1.22}$ **K**

(e)   When $\alpha = 1.2 \times 10^{-6}$ K$^{-1} \rightarrow \delta T = \mathbf{19.3}$ **K**

END OF SECTION

Mathematics for Physics

Question 1

We need to realise that $a^2 - b^2$ is equal to $(a - b)(a + b)$. Hence, $6667^2 - 3333^2 = (6667 - 3333)(6667 + 3333) = (3334)(10{,}000) = \mathbf{33{,}340{,}000}$.

Question 2

To find the gradient of a tangent to a curve, we need to differentiate the curve equation, $\frac{dy}{dx} = m = 4x^3$. At (-2,16), $m = 4(-2)^3 = -32$. To find the equation of the tangent, we can use $(y - y_1) = m(x - x_1)$. Substituting the coordinate point, we get $(y - 16) = -32(x - (-2))$; $\boldsymbol{y = -32x - 48}$

Question 3

We can simplify the expression $\frac{2log\ 125}{3log\ 25}$ $into$ $\frac{2log\ 5^3}{3log\ 5^2} = \frac{(3)2log\ 5}{(2)3log\ 5} = \frac{6}{6} = \mathbf{1}$

Question 4

(i)   To add up to 6, there are several combination possibilities: (1+5), (2+4), (3+3), (4+2), and (5+1). Each of this possibility has a probability of $\frac{1}{6} * \frac{1}{6} = \frac{1}{36}$. Adding up all possibilities, then the total probability $P = 5 * (\frac{1}{36}) = \frac{5}{36}$

(ii)   Possible scenarios for 11 as total: (5+6) and (6+5). Total probability $P = 2 * (\frac{1}{36}) = \frac{1}{18}$

Question 5

This is binomial expansion and you need to be familiar with this (along with *Pascal's Triangle*). The expansion of $(2 + x)^5$ is $2^5 + (5)(2)^4x + (10)(2)^3(x)^2 + (10)(2)^2(x)^3 + \cdots = \mathbf{32 + 80x + 80x^2 + 40x^3 + \cdots}$

Question 6

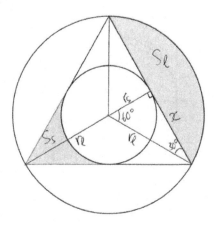

from the above sketch, we can deduce that $sin\ 30^o = \frac{1}{2} = \frac{r_s}{r_l}$; $r_l = 2r_s$

i) $\quad \frac{A_l}{A_s} = \frac{\pi r_l^2}{\pi r_s^2} = \frac{(2r_s)^2}{r_s^2} = \mathbf{4}$

ii) $\quad$ Height of triangle $= r_l + r_s$; $x = r_l cos\ 30^o = \frac{\sqrt{3}r_l}{2}$

Area of triangle $= \frac{1}{2}(2x)(r_l + r_s) = \frac{\sqrt{3}r_l(r_l+r_s)}{2} = 3\sqrt{3}r_s^2 = k$

From the sketch above, we know that: $S_s = \frac{area\ of\ triangle - area\ of\ small\ circle}{3} = \frac{k - A_s}{3}$; and

$S_l = \frac{area\ of\ large\ circle - area\ of\ triangle}{3} = \frac{A_l - k}{3}$

Hence $\rightarrow \frac{S_l}{S_s} = \frac{A_l - k}{k - A_s} = \frac{4\pi r_s^2 - 3\sqrt{3}r_s^2}{3\sqrt{3}r_s^2 - \pi r_s^2} = \frac{4\pi - 3\sqrt{3}}{3\sqrt{3} - \pi}$

Question 7

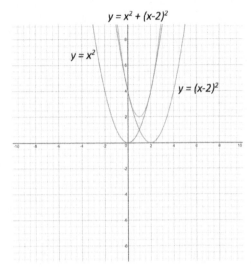

$y = x^2$ graph should be quite generic, and you should be familiar with it. $y = (x - 2)^2$ is $y = x^2$ curve, but shifted by 2 points along the *x-axis* (blue curve). $y = x^2 + (x - 2)^2$ is the sum of the first two curves, hence giving us the green curve.

Question 8

$tan \, \theta \, = 2sin \, \theta$; but we know that $tan \, \theta \, = \frac{sin \, \theta}{cos \, \theta}$ so $\frac{sin \, \theta}{cos \, \theta} = 2sin \, \theta$

When $sin \, \theta \, = 0$; then $\theta = 0 \, , \pi \, , 2\pi$

When $in \, \theta \, \neq 0$; then $\frac{sin \, \theta}{cos \, \theta} = 2sin \, \theta$; dividing both sides by $sin\theta$ gives $\frac{1}{cos \, \theta} = 2$; $cos \, \theta \, = \frac{1}{2} \rightarrow \theta = \frac{\pi}{3}$ **or** $\theta = \frac{5\pi}{3}$

Question 9

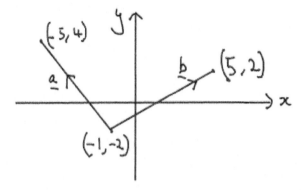

From the three points, we can construct two vectors as shown above. Then $\underline{a} = \binom{-5}{4} - \binom{-1}{-2} = \binom{-4}{6}$ and $\underline{b} = \binom{5}{2} - \binom{-1}{-2} = \binom{6}{4}$

$[\underline{a}] = \sqrt{(-4)^2 + 6^2} = \sqrt{52}$

$[\underline{b}] = \sqrt{6^2 + 4^2} = \sqrt{52} = [\underline{a}]$

$\underline{a} \cdot \underline{b} = \binom{-4}{6} \cdot \binom{6}{4} = 0$

∴The sides are perpendicular and have equal length.

The 4th corner is $\left(\frac{-5}{4}\right) + \left(\frac{6}{4}\right) = \left(\frac{1}{8}\right)$ and area $= (\sqrt{52})^2 = \mathbf{52}$

Question 10

$$\int_1^9 \sqrt{x} + \frac{1}{\sqrt{x}} \, dx = \int_1^9 x^{\frac{1}{2}} + x^{\frac{-1}{2}} dx = \left| \frac{2}{3} x^{\frac{3}{2}} + 2x^{\frac{1}{2}} \right|_1^9$$

$$= \left(\frac{2}{3}(9)^{\frac{3}{2}} + 2(9)^{\frac{1}{2}}\right) - \left(\frac{2}{3}(1)^{\frac{3}{2}} + 2(1)^{\frac{1}{2}}\right) = \frac{\mathbf{64}}{\mathbf{3}}$$

Question 11

From the statements in the question we can deduce:

$$ar = a + d \qquad \rightarrow d = a(r-1) \qquad (1)$$

$$ar^2 = 2(a + 2d) \rightarrow d = \frac{a(r^2-2)}{4} \qquad (2)$$

Equating these two, we get:

$$a(r-1) = \frac{a(r^2-2)}{4}$$

$$4r - 4 = r^2 - 2$$

$$r^2 - 4r + 2 = 0$$

Solving this quadratic expression using $x = \frac{-b \pm \sqrt{b^2 - 4ac}}{2a}$, we get $\boldsymbol{r_{1,2} = 2 \pm \sqrt{2}}$

Question 12

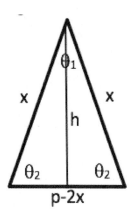

Pythagoras: $h^2 + (\frac{p-2x}{2})^2 = x^2$

$$h = (x^2 - (\frac{p-2x}{2})^2)^{\frac{1}{2}} = (\frac{1}{4}(4px - p^2))^{\frac{1}{2}}$$

$$Area = A = \frac{1}{2} * base * height = \frac{1}{2}(p - 2x)(h) = \frac{1}{2}(p - 2x)(\frac{1}{4}(4px - p^2))^{\frac{1}{2}}$$

$$= \frac{1}{4}(p - 2x)(4px - p^2)^{\frac{1}{2}}$$

To maximise the area, $\frac{dA}{dx} = 0$

$$\rightarrow \frac{dA}{dx} = \frac{1}{4}((-2)(4px - p^2)^{\frac{1}{2}} + (p - 2x)(\frac{1}{2}(4px - p^2)^{\frac{-1}{2}}(4p))) = 0$$

$$p(p - 2x) = 4px - p^2 \rightarrow x = \frac{p}{3}$$

$$cos\,\theta_2 = \frac{\frac{1}{2}(p - \frac{2p}{3})}{\frac{p}{3}} = \frac{1}{2} \rightarrow \theta_2 = 60^o$$

So, $\theta_1 = 180^o - 2\theta_2 = 60^o$

END OF PAPER

2008

Part A

Question 1

1+2+3+...+99+100 is an arithmetic series with first term 1 and common difference 1. So,

$$S_n = \frac{n}{2}(a_1 + a_n)$$

$$S_{100} = \frac{100}{2}(1 + 100) = 50(101) = \mathbf{5050}$$

Question 2

$$(0.25)^{-\frac{1}{2}} = (\frac{1}{4})^{-\frac{1}{2}} = (4)^{\frac{1}{2}} = \mathbf{2}$$

$$(0.09)^{\frac{3}{2}} = (\frac{9}{100})^{\frac{3}{2}} = (\frac{3}{10})^3 = \frac{27}{1000} = \mathbf{0.027}$$

Question 3

We know that the first three terms of $(1 + a)^n \approx 1 + na + \frac{n(n-1)a^2}{2} + \cdots$..

For $(1 + x)^{m+1}$; $a = x$ and $n = m + 1$

For $(1 - 2x)^m$; $a = -2x$ and $n = m$

So, $(1 + x)^{m+1}(1 - 2x)^m \approx (1 + (m + 1)x + \frac{(m+1)mx^2}{2})(1 + m(-2x) + \frac{m(m-1)(-2x)^2}{2})$;

$$\approx \mathbf{1 + (1 - m)x + (\frac{m^2 - 7m}{2})x^2}$$

Question 4

$\frac{x^2+2}{1-x^2} < 3$; we know that $1 - x^2 \neq 0$; (because otherwise the left-hand side becomes infinity) $\rightarrow x \neq \pm 1$

$x^2 + 2 < 3(1 - x^2) \rightarrow 4x^2 < 1$;

$$x^2 - \frac{1}{4} < 0 ; \rightarrow \left(x - \frac{1}{2}\right)\left(x + \frac{1}{2}\right) < 0;$$

$$x = \pm\frac{1}{2}$$

Then, we make a number line with points of interest 1, -1, ½ and -1/2.

Hence the answer is $\frac{-1}{2} < x < \frac{1}{2}$ **or** $x < -1$ or $x > 1$

Question 5

i) $\log_2 9 = \frac{\log_9 9}{\log_9 2} = \frac{1}{x}$

ii) $\log_8 3 = \frac{\log_9 3}{\log_9 8} = \frac{1/2}{\log_9(2)^3} = \frac{1/2}{3\log_9 2} = \frac{1/2}{3x} = \frac{1}{6x}$

Question 6

For $1, x^2, x$ to be successive terms of an arithmetic progression, then:

$x^2 = 1 + d ; d = x^2 - 1$ (1)

$x = 1 + 2d$ (2)

Substituting (1) into (2) we get $x = 1 + 2(x^2 - 1) \rightarrow 2x^2 - x - 1 = 0$

$(2x + 1)(x - 1) = 0 \rightarrow x = \frac{-1}{2}$ or $x = 1$

Question 7

The expression for the tangent of curve y is $\frac{dy}{dx} = 1 + x + x^2 + x^3 + \cdots$

At $x = 0$, $\frac{dy}{dx} = 1$ hence $a = 1$ (since the gradient of the line is always a at any point).

At $x = \frac{1}{4}$, $\frac{dy}{dx}$ will be a geometric progression with $r = x = \frac{1}{4}$. $\frac{dy}{dx} = S_\infty = \frac{u_1}{1-r} = \frac{1}{3/4} = \frac{4}{3}$; hence $a = \frac{4}{3}$.

Question 8

Centre of the circle (x,y) will be the mid-point of the opposite ends $= \left(\frac{5+(-3)}{2}, \frac{2+8}{2}\right) = (1,5)$

Diameter of the circle $= \sqrt{(5-(-3))^2 + (2-8)^2} = 10$; radius $= 5$

Hence, the equation of the circle $= (x-1)^2 + (y-5)^2 = 25$

Question 9

$P(x=5 \text{ or } 6) = 3a$

$P(x=2 \text{ or } 3 \text{ or } 4) = a$

$P(x=1) = 0.$

$P(\text{total}) = 1 = \sum_{i=1}^{6} P(x = n) = 0 + a + a + a + 3a + 3a = 9a \rightarrow a = \frac{1}{9}$

   i) $P(x=2) = a = \frac{1}{9}$

   ii)   To get a total of ≥ 10, the possible combinations are $(4,6)$, $(5,5)$, $(6,4)$, $(5,6)$, $(6,5)$ or $(6,6)$

$$P = (a)(3a) + (3a)(3a) + (3a)(a) + (3a)(3a) + (3a)(3a)$$
$$+ (3a)(3a) = 42a^2 = \frac{14}{27}$$

Question 10

Body diagonal of a cube is $\sqrt{3}side = \sqrt{3}\mathbf{a}$

Question 11

   i) $\int_{-1}^{1} x + x^3 + x^5 + x^7 dx = [\frac{1}{2}x^2 + \frac{1}{4}x^4 + \frac{1}{6}x^6 + \frac{1}{8}x^8]_{-1}^{1} =$

$\left(\frac{1}{2} + \frac{1}{4} + \frac{1}{6} + \frac{1}{8}\right) - \left(\frac{1}{2} + \frac{1}{4} + \frac{1}{6} + \frac{1}{8}\right) = 0$

   ii) $\int_{0}^{1} \frac{x^9 + x^{99}}{11} dx = [\frac{1}{110}x^{10} + \frac{1}{1100}x^{100}]_{0}^{1} = \left(\frac{1}{110} + \frac{1}{1100}\right) = \frac{11}{1100} = \frac{1}{100}$

Question 12

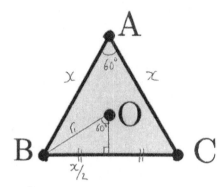

$$A_{ABC} = \frac{1}{2}x^2 \sin 60^o = \frac{\sqrt{3}}{4}x^2$$

$$\sin 60^o = \frac{x/2}{r_1} \; ; \; r_1 = \frac{x}{\sqrt{3}}$$

Area of small circle, $A_1 = \pi r_1^2 = \frac{\pi x^2}{3}$

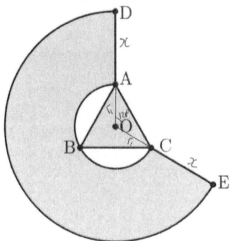

Area of big circle $A_2 = \pi(x + r_1)^2 = \pi x^2 + \frac{\pi x^2}{3} + \frac{2\pi x^2}{\sqrt{3}} = 2\pi x^2 \left(\frac{2+\sqrt{3}}{3}\right)$

$$A_{ADEC} = \frac{240}{360}(A_2 - A_1) = \frac{2}{3}\left(2\pi x^2 \left(\frac{2+\sqrt{3}}{3}\right) - \frac{\pi x^2}{3}\right) = \frac{2\pi x^2}{9}(3 + 2\sqrt{3})$$

$$\frac{A_{ADEC}}{A_{ABC}} = \frac{2\pi x^2}{9}(3 + 2\sqrt{3}) \div \frac{\sqrt{3}}{4}x^2 = \frac{8\pi(\sqrt{3} + 2)}{9}$$

END OF SECTION

Part B

Question 13: D

The moments have to be equal at both sides, ie. $(m_1g)l_1 = (m_2g)l_2$ with l calculated from the centre of the seesaw.

$$l_2 = \frac{m_1 l_1}{m_2} = \frac{(20)(1.5)}{30} = 1m$$

Hence, the girl has to be $1 + 1.5 =$ **2.5 m away from the boy**

Question 14: B

When energy is released, mass has to decrease since the energy comes from the mass defect.

Question 15: D

Mass of stars in the universe $= 2 * 10^{30} * 250 * 10^9 * 400 * 10^9 = 2 * 10^{53}\ kg$

Mass of dark matter $= 20 * mass\ of\ stars = 4 * 10^{54}\ kg$

Total mass of the universe = **4.2 * 10⁵⁴ kg**

Question 16: A

Solar eclipse occurs when the moon is between earth and the sun, hence new moon.

Question 17: C

The pressure will rise as the gas molecules will hit the container wall more intensely and with more energy; the density remains the same as the volume is fixed and so the mass (there are no loss or gain of matters).

Question 18: B

The minimum energy required is the potential energy needed to climb the height, $GPE = mgh = 60 * 10 * 4 =$ **2400 J**.

Question 19: C

Energy in the battery $= V * i * t = 3.6 * 0.7 * 3600 = 9072\ J$

Power that can be harvested using the solar panel $= (0.25)^2 * 1000 * 0.1 = 6.25\ W$

Time taken to charge $= \frac{E}{P} = \frac{9072}{6.25} = 1451\ s =$ **0.4 hours**

Question 20: C

Power is dissipated highest when the resistance is lowest $(P = \frac{V^2}{R})\rightarrow$ when the light intensity is highest. This happens during noon.

Question 21: A

It can be assumed that all kinetic energy is converted into heat $\rightarrow E_k = \frac{1}{2}mv^2 = \frac{1}{2}(0.01)(400)^2 = 800J$

Heat energy $= mc\Delta T = \rho Vc\Delta T = \rho(s)^3 c\Delta T;\ \Delta T = \frac{800}{1000(2)^3 4200} = \mathbf{2.4\ x\ 10^{-5}K}$

Question 22: A

Since the unit used is gram which is 10^{-3} kg, then the units for energy would therefore be 10^{-3} J which is mJ.

Question 23

Given that $q = CV$; $C = \frac{\rho A}{d}$

$$W = \frac{q^2}{2C} = \frac{C^2V^2}{2C} = \frac{CV^2}{2} = \frac{\rho AV^2}{2d}$$

Given that $V_{max} = Bd$; $E_{max} = \frac{\rho AV_{max}^2}{2d} = \frac{\rho AB^2 d^2}{2d} = \frac{\rho AB^2 d}{2} = \frac{\rho B^2 V}{2}$

Using $m = DV$; $E_{max} = \frac{mpB^2}{2D}$

For $p = 2x10^{-11}Fm^{-1}, B = 2x10^7 Vm^{-1}, D = 1000\ kgm^{-3}, m = 1kg;$

$E_{max} = \mathbf{4J}$

1 kW = 1000 J/s → 4 J capacity is too small to have a useful effect

Question 24

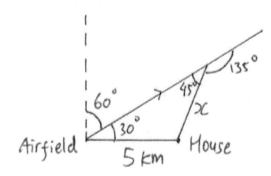

$$\frac{x}{\sin 30} = \frac{5}{\sin 45} \; ; x = \frac{5 \sin 30}{\sin 45} = 2.5\sqrt{2} = \textbf{3.5 } \textit{\textbf{km}}$$

Question 25

Statement (a) → $c + r = 2s \; ; r = 2s - c$ (1)
Statement (b) → $c^2 + s^2 = r^2$ (2)
Statement (c) → $2c + s = 1 \; ; s = 1 - 2c$ (3)

Substitute (3) into (1) → $r = 2(1 - 2c) - c = 2 - 5c$
Then this into (2) → $c^2 + (1 - 2c)^2 = (2 - 5c)^2$
$c^2 + 1 - 4c + 4c^2 = 4 - 20c + 25c^2 \rightarrow 20c^2 - 16c + 3 = 0$
$(10c - 3)(2c - 1) = 0 \; ; c = \frac{3}{10}$ or $\frac{1}{2}$; but from (3), c cannot be ½ as this would
give $s = 0$.
$\therefore \textbf{\textit{c}} = \textbf{0.3} \textit{\textbf{m}}, \textbf{\textit{s}} = \textbf{0.4} \textit{\textbf{m}}, \textbf{\textit{r}} = \textbf{0.5} \textit{\textbf{m}}$

Question 26

Most rapid change is when the gradient is the steepest → **7 am.**

$$m = \frac{\Delta h}{\Delta t} = \frac{250 - 50}{105 - 15}$$
$$= \textbf{2.2 } \textit{\textbf{cm}}$$
$$\textbf{\textit{/min}}$$

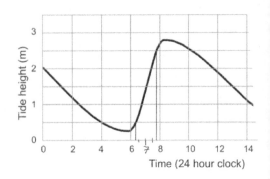

Question 27

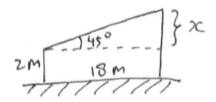

$\tan 45^o = \dfrac{x}{18} = 1 \; ; x = \mathbf{18m}$

a) $s = 20m, u = 0, v = x, a = 10ms^{-2}, t =?$

$s = ut + \dfrac{1}{2}at^2;$

$20 = 5t^2 \; ; \mathbf{t = 2s}$

b) $v^2 = u^2 + 2aS$

$v = \sqrt{2*10*20} = \mathbf{20ms^{-1}}$

c) $\frac{1}{2}mv^2 = Fd$

$F = \dfrac{\frac{1}{2}*0.02*20^2}{1*10^{-3}} = \mathbf{4000N = 4kN}$

d) $WD = Fd = 4*10^3 * 10^{-3} = \mathbf{4J}$

$GPE = mgh = 0.02*10*20 = \mathbf{4J = WD}$

e) $d = 5cm \rightarrow F = \dfrac{0.5*0.02*20^2}{0.05} = \mathbf{80N}$

$Ft = m(v - u)$

$t = \dfrac{0.02(0-20)}{-80} = \dfrac{1}{200} = \mathbf{5\ ms}$

f) $E_{min} = mgh = (100 + 0.02)*10*20 = \mathbf{2.0*10^4\,J}$

g) $E_{min} * Efficiency = mc\Delta T$

$Efficiency = \dfrac{0.02*4000*80}{2*10^4} = 0.32 = \mathbf{32\%}$

END OF PAPER

2009

Part A:

Question 1

$$y = \tan t = \frac{\sin t}{\cos t} = \frac{\sin t}{\sqrt{1 - \sin^2 t}} = \frac{x}{\sqrt{1 - x^2}}$$

Question 2

$y = x + \frac{4}{x^2}$ can be constructed from two separate

functions $y = x$ and $y = \frac{4}{x^2}$

You should be familiar with how these two graphs look like.

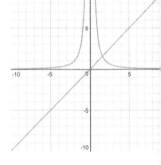

To find stationary point, we need to differentiate the

whole function $\rightarrow \frac{dy}{dx} = 1 + (-2)\frac{4}{x^3} = 1 - \frac{8}{x^3} = 0$

$x^3 = 8 \rightarrow x = 2$

$y = (2) + \frac{4}{(2)^2} = 3 \rightarrow stationary\ point = (2, 3)$

When $x = 4$, $y = 4.25$ → (4, 4.25)
When $x = -4$, $y = -3.75$ → (-4, -3.75)

Combining the two curves, having known these three points, will give you the curve below:

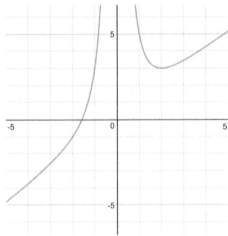

Question 3

We can sketch the diagram of the lines and the circle as follow:

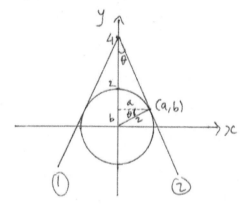

Looking at the two θ's drawn, we can say:

$\sin \theta = \dfrac{2}{4} = \dfrac{b}{2} \to b = 1$

Look at the small triangle, and we can say $b^2 + a^2 = 2^2 = 4 \to a = \sqrt{4 - b^2} = \sqrt{4 - 1} = \sqrt{3}$

We then can work out the gradient of line 2 $\to m_2 = \dfrac{y_2 - y_1}{x_2 - x_1} = \dfrac{1-4}{\sqrt{3}-0} = \dfrac{-3}{\sqrt{3}} = -\sqrt{3}$

We then know that the gradient of line 1 has to be $\sqrt{3}$.

$\therefore y_1 = \sqrt{3}x + 4 \ and \ y_2 = -\sqrt{3}x + 4$

Question 4

i) $\log_2 \sqrt[3]{x} = \dfrac{1}{2} \to 2^{\frac{1}{2}} = \sqrt[3]{x}$

$\therefore x = (2^{\frac{1}{2}})^3 = 2^{\frac{3}{2}} = 2\sqrt{2}$

ii) $\sqrt{\log_8 16} = \sqrt{\log_8 (8)^{\frac{4}{3}}} = \sqrt{\dfrac{4}{3} \log_8 8} = \sqrt{\dfrac{4}{3}} = \dfrac{2}{\sqrt{3}}$

Question 5

We can introduce a substitute variable $y = x^2$, then our equation becomes:

$y^2 - 13y + 36 = 0$

$(y - 9)(y - 4) = 0$

$y = 9 \ or \ 4 = x^2$

$\therefore x = \pm 2 \ or \ \pm 3$

Question 6

The right-hand side of the inequality is geometric progression with common ratio $\frac{1}{x}$. The sum to infinite term is then $S_\infty = \frac{u_1}{1-r} = \frac{1}{1-\frac{1}{x}} = \frac{x}{x-1}$

Our inequality then becomes:

$$x + 2 < \frac{x}{x-1}$$
$$\rightarrow (x+2)(x-1)^2 < x(x-1)$$
$$\rightarrow x^3 - 3x + 2 < x^2 - x$$
$$\rightarrow x^3 - x^2 - 2x + 2 < 0$$
$$\rightarrow (x-1)(x^2 - 2) < 0$$
$$\rightarrow (x-1)(x+\sqrt{2})(x-\sqrt{2}) < 0$$

We then make the number line with points of interest $1, -\sqrt{2} \text{ and } \sqrt{2}$.

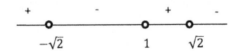

but given that $x > 0$; then $\therefore \mathbf{1 < x < \sqrt{2}}$

Question 7

The possible combination to give a total of 10 or higher is (4,6), (6,4), (5,5), (5,6), (6,5) and (6,6) and each scenario has a probability of $\frac{1}{36}$. Hence, the total probability is $6 * \frac{1}{36} = \frac{1}{6}$

For the third die to bring a total of 15 or higher, then:
→ If the total for the first two is 10 (Probability of this is 3/6 (from above)) → the third die has to be 5 or 6 (probability of this is 2/6) → total probability of 3/6 * 2/6 = 6/36
→ If the total for the first two is 11 (probability 2/6)→ the third die has to be 4, 5, or 6 (probability 3/6)→ total probability of 2/6 * 3/6 = 6/36
→ If the total for the first two is 12 (probability 1/6) → the third die can be either 3, 4, 5 or 6 (probability 4/6) → total probability of 1/6 * 4/6 = 4/36

$$\therefore P_{tot} = \frac{6}{36} + \frac{6}{36} + \frac{4}{36} = \frac{16}{36} = \frac{4}{9}$$

Question 8

We can substitute $1 - 2sin^2x = \cos 2x$ (remember the sine and cosine rules).
Hence, the sketch of $y = \cos 2x$ will just be a cos x function but compressed in the x-axis by a factor of two.

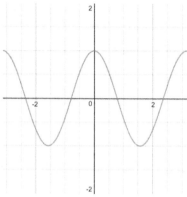

Question 9

First, we need to find the limits of the area (x points where $y = 0$)

$$y = \frac{(x^2 - 4x)}{\sqrt{x}} = 0$$

$$x^2 - 4x = 0$$

$$x(x - 4) = 0 \rightarrow x = 0 \; or \; 4$$

The area is bound between the x-axis ($y = 0$ line) and the curve, hence:

$$A = \int_0^4 0 - \left(\frac{x^2 - 4x}{\sqrt{x}}\right) dx = -\int_0^4 x^{\frac{3}{2}} - 4x^{\frac{1}{2}} dx = -[\frac{2}{5}x^{\frac{5}{2}} - \frac{8}{3}x^{\frac{3}{2}}]_0^4$$

$$= -\left(\frac{2}{5}(4)^{\frac{5}{2}} - \frac{8}{3}(4)^{\frac{3}{2}}\right) = \frac{128}{15}$$

Question 10

We can sketch the figure as below:

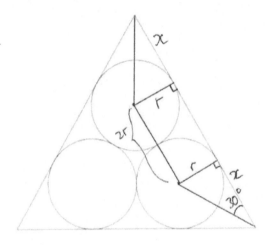

From above, we can see that $\tan 30^o = \frac{r}{x} = \frac{\sqrt{3}}{3} \rightarrow x = \sqrt{3}r$

Hence, the side of triangle $= x + 2r + x = \left(2 + 2\sqrt{3}\right)r$

Area of triangle $= \frac{1}{2}(side)^2 \sin 60^o = \frac{1}{2}\left(2 + 2\sqrt{3}\right)^2 r^2 \frac{\sqrt{3}}{2} = \left(4\sqrt{3} + 6\right)r^2$

$\therefore \frac{A_{circles}}{A_{triangle}} = \frac{3\pi r^2}{\left(4\sqrt{3} + 6\right)r^2} = \frac{3\pi}{\left(4\sqrt{3} + 6\right)}$

Question 11

If they are an arithmetic progression, then they will have a common difference between them. Hence:

$a - \frac{a}{b} = \frac{a}{b} - \left(-\frac{a}{b}\right)$

$a = \frac{3a}{b}$

$\therefore \boldsymbol{b = 3}$

Question 12

We can do binomial expansion to solve this.

$(2.1)^5 = (2 + 0.1)^5 = 2^5 + 5(2)^4(0.1) + 10(2)^3(0.1)^2 + 10(2)^2(0.1)^3 + \cdots$

$= 32 + 8 + 0.8 + 0.04 = \boldsymbol{40.8}$

END OF SECTION

Part B

Question 13: A
To answer this question we need to apply mass-energy conservation $E = mc^2$

For every second, $E = mc^2 = 3.8 * 10^{26} J$

$$m = \frac{3.8 * 10^{26}}{(3 * 10^8)^2} = \textbf{4.2} * \textbf{10}^9 \textbf{ kg}$$

Question 14: C
As the battery is connected in parallel, the maximum voltage stays the same. However, current will be added up hence maximum current is higher.

Question 15: C
A total solar eclipse cannot be seen since the angular diameter of the sun is larger than the angular diameter of the titan, as seen from the surface of Saturn.

Question 16: C
This is Archimedes' Principle question. Initially, the boat (with the its anchor) floats on the lake, hence $(M + m)g = V_d \rho_w g$, with M=mass of boat, m=mass of anchor, V_d is volume of water displaced and ρ_w being the water density. From this, we get that $V_d = \frac{M+m}{\rho_w}$.

After the anchor is dropped, the total displaced volume consists of volume displaced by the boat and volume displaced by the anchor. The volume displaced by the boat is $\frac{M}{\rho_w}$, and the volume displaced by the anchor is equal to the anchor volume, as it is fully submerged/sink. Hence, the final displaced volume is $V_d' = \frac{M}{\rho_w} + V_{anchor} = \frac{M}{\rho_w} + \frac{m}{\rho_{anchor}}$.

As $\rho_{anchor} > \rho_w$, then V_d' must be less than V_d (the final displaced volume is less than that of initial before the anchor is dropped). **Hence the water level will fall slightly**.

Question 17: B
In this case potential energy can be assumed to be 100% converted into thermal energy. So:

$$mgh = mc\Delta T$$

$$\Delta T = \frac{gh}{c} = \frac{10 * 105}{4200} = \textbf{0.25}^o \textbf{C}$$

Question 18: D

Rearrange the equation to give $m = 2qU\left(\frac{t}{d}\right)^2 = 2(1.6 *$

$10^{-19})(16000)\left(\frac{30*10^{-6}}{1.5}\right)^2 = 2 * 10^{-24}kg$

Question 19: D

$K = \dfrac{F}{v^2 A} = \dfrac{kgms^{-2}}{m^2 s^{-2}.m^2} = kgm^{-3}$

$\therefore K$ is **density**

Question 20: C

One of the resistors is replaced hence resistance decreases so I initially increase. But as the capacitor is being charged up, the current decreases.

Question 21: B

As light is moving into a more dense medium, the refraction will go towards normal hence it will bend down.

Question 22: A

Linear acceleration can be calculated as $a = \omega^2 r = \left(\frac{2\pi}{T}\right)^2 r = \frac{4\pi^2(400,000*10^3)}{(2.4*10^6)^2} =$

$2.7 * 10^{-3}ms^{-2}$

Question 23

We rearrange the equation to $A = \dfrac{P}{kT^4} = \dfrac{75}{(6*10^{-8})(5000)^4} = 2 * 10^{-6}m^2$

Question 24

Let p, q and r equal to the lengths of pangs, quizzers and roodles respectively. Then:

$$q + r = 2p \; ; q = 2p - r \qquad (1)$$
$$q^2 = 2p^2 + r^2 \qquad (2)$$

Substitute (1) into (2):

$$(2p - r)^2 = 2p^2 + r^2$$
$$\rightarrow 4p^2 - 4pr + r^2 = 2p^2 + r^2$$
$$\rightarrow 2p^2 = 4pr$$
$$\therefore p = 2r \qquad (3)$$

Substitute (3) into (1):

$$q = 2p - r = 2(2r) - r = 3r$$

Let $q^3 = ap^3 + br^3$

Where a = no. of pangs used
          b = no. of roodles used

$$\rightarrow (3r)^3 = a(2r)^3 + br^3$$
$$\rightarrow 27r^3 = 8ar^3 + br^3$$
$$\therefore 27 = 8a + b \; ; \boldsymbol{b = 27 - 8a} \qquad \textbf{(4)}$$

We need to minimise $ap^2 + br^2 = (4a + b)r^2$

Possible solutions are:

a	b (from (4))	4a+b
1	19	23
2	11	19
3	3	15

$\therefore \boldsymbol{pangs = 3, roodles = 3}$

Question 25

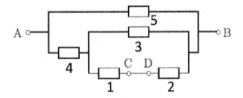

All resistors are $1k\Omega$.

$$R_{1-3} = \frac{1}{\frac{1}{1} + \frac{1}{2}} = \frac{2}{3} k\Omega$$

$$R_{1-4} = \frac{2}{3} + 1 = \frac{5}{3} k\Omega$$

$$R_{1-5} = \frac{1}{\frac{1}{1} + \frac{3}{5}} = \frac{5}{8} k\Omega = \mathbf{625\Omega}$$

When battery is connected, $R_{Tot} = 625 + 125 = 750\Omega$

$$V_{AB} = \frac{625}{750} V_{Tot} = \frac{625}{750} (6) = 5V = V_{1-4}$$

$$V_{1-3} = V_{1-4} * \frac{R_{1-3}}{R_{1-4}} = 5 * \frac{\frac{2}{3}}{\frac{5}{3}} = 2V = V_{1-2}$$

$$I_{CD} = \frac{V_{1-2}}{R_{1-2}} = \frac{2}{(1+1) * 10^3} = \mathbf{1\ mA}$$

Question 26

We know that kinetic energy equals electrical potential energy, $\frac{1}{2} mv^2 = eV$

$$v^2 = \frac{2eV}{m} = \frac{2 * 1.6 * 10^{-19} * 50}{10^{-30}} = 1.6 * 10^{13};$$

$v = 4 * 10^6 \frac{m}{s} \rightarrow$ this is the horizontal velocity from the electron gun

Time taken to reach screen: $t = \frac{d}{v} = \frac{0.4}{4*10^6} = 10^{-7}\ s$

Due to gravity, the electron will "fall" downward as well. The total fall can be calculated as: $S = ut + 1/2at^2$

Initial downward velocity u is zero, and a is gravitational acceleration = 10 m/s^2

$$\therefore S = \frac{1}{2} (10)(10^{-7})^2 = \mathbf{5 * 10^{-14}\ m}$$

Question 27

a) The energy dispersed by the brakes will equal the kinetic energy of the car:

$$\therefore E_{brakes} = \frac{1}{2}mv^2$$

b) Time between subsequent stops is $t = \frac{s}{v}$, average power dissipated $P =$

$$\frac{E}{t} = \frac{1}{2}mv^2 * \frac{v}{s} = \frac{mv^3}{2s}$$

c) For traveling a total distance d, the total time taken is $t = \frac{d}{v}$.

$$E = Pt = P * \frac{d}{v} = \frac{mv^3}{2s} * \frac{d}{v} = \frac{mv^2 d}{2s}$$

d) $E_{10} = \frac{(1000)(10)^2(1000)}{2(100)} = \mathbf{5 * 10^5 \, J}$

If speed is doubled:

$E_{20} = 4E_{10} = \mathbf{2 * 10^6 \, J}$ (4 times since $E \propto v^2$)

e)

The car will travel horizontally as far as vt. Hence, the volume of air swept $V = Avt$

$E_{air} = \frac{1}{2}mv^2 = \frac{1}{2}(DV)v^2 = \frac{1}{2}DAtv^3$ (remember that mass = density * volume)

$$P = \frac{E_{air}}{t} = \frac{1}{2}DAv^3$$

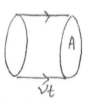

f) $E_{tot} = Energy \; for \; travel + Energy \; transferred \; to \; air = E + E_{air}$

$$\therefore E_{tot} = 5 * 10^5 + \frac{1*1*\left(\frac{1000}{10}\right)*(10)^3}{2} = \mathbf{5.5 * 10^5 \, J}$$ (remember that $t = \frac{d}{v}$ and E is from part (d))

g) Between stops, $s = vt$ (t here is the time between stop). Hence:

$E_{brakes} = E_{air}$

$$\rightarrow \frac{1}{2}mv^2 = \frac{1}{2}DAtv^3 = \frac{1}{2}DAsv^2$$

$$\therefore s = \frac{m}{DA} = \frac{1000}{(1)(1)} = \mathbf{1000 \, m}$$

h) On highways, s will be large, more energy is lost to air resistance. Hence smaller A is beneficial.

In cities, s is small and more energy is lost to braking. Hence lighter designs are crucial → smaller mass.

END OF PAPER

2010

Part A:

Question 1

   i) $\sin 3x = \sqrt{3}\cos 3x \; ; \rightarrow \; \tan 3x = \sqrt{3}$

       Hence, $3x = \dfrac{\pi}{3}, \dfrac{4\pi}{3}, \dfrac{7\pi}{3}$

       $\therefore x = \dfrac{\pi}{9}, \dfrac{4\pi}{9}, \dfrac{7\pi}{9}$

ii) using $sin^2 x + cos^2 x = 1$ identity:

$1 - sin^2 x - sinx + 1 = 0$

$sin^2 x + sinx - 2 = 0$

$(sinx + 2)(sinx - 1) = 0$

Either $sinx = -2 \; (no \; solution) or \; sinx = 1$

$\therefore x = \dfrac{\pi}{2}$

Question 2

The smaller circle has diameter of the radius of the bigger circle → $r = 0.5$

From the sketch, we can tell that the centre of the smaller circle M is $(1.5, 1.5)$

$\therefore \left(y - \dfrac{3}{2}\right)^2 + \left(x - \dfrac{3}{2}\right)^2 = \dfrac{1}{4}$

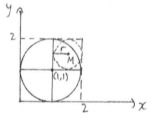

Question 3

If $x = -1$ is a root, then:

$(-1)^3 + 2(-1)^2 - 5(-1) - 6 = -1 + 2 + 5 - 6 = 0$

$\therefore x = -1 \; is \; a \; root$

$x^3 + 2x^2 - 5x - 6 = (x + 1)(x^2 + x - 6) = 0$

$(x + 1)(x + 3)(x - 2) = 0$

$x = -1, -3 \; or \; 2$

Question 4

$$y - y_1 = m(x - x_1) \qquad \text{and } m = \frac{y_2 - y_1}{x_2 - x_1}$$

$$y - 3 = \left(\frac{5-3}{1-2}\right)(x - 2) = -2x + 4$$

$$\therefore y = -2x + 7$$

Question 5

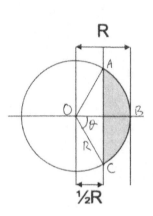

$$\cos \theta = \frac{1/2 R}{R} = \frac{1}{2} \rightarrow \theta = 60^o$$

$$\text{Area of OABC} = \frac{120}{360}\pi R^2 = \frac{\pi R^2}{3}$$

$$\text{Area of } \Delta OAC = \frac{1}{2}R^2 \sin 120 = \frac{\sqrt{3}R^2}{4}$$

$$\therefore \textbf{shaded area} = \frac{\pi R^2}{3} - \frac{\sqrt{3}R^2}{4} = R^2\left(\frac{\pi}{3} - \frac{\sqrt{3}}{4}\right)$$

Question 6

$$2x + 2y = L; \rightarrow y = \frac{L}{2} - x$$

$$Area = xy = x\left(\frac{L}{2} - x\right) = \frac{L}{2}x - x^2$$

To maximise area,

$$\frac{dArea}{dx} = 0 = \frac{L}{2} - 2x; \rightarrow x = \frac{L}{4}$$

$$\therefore Area\ max. = xy = \frac{L}{2}\left(\frac{L}{4}\right) - \left(\frac{L}{4}\right)^2 = \frac{L^2}{16}$$

Question 7

i. This question can be answered straightaway. $\log_3 9 = 2$

ii. $\log 4 + \log 16 - \log 2 = \log 2^2 + \log 2^4 - \log 2 = 2\log 2 + 4\log 2 - \log 2 = \textbf{5} \log \textbf{2}$

Question 8

i. $(16.1)^2 = (16 + 0.1)^2 = 16^2 + 2(16)(0.1) + 0.1^2$
$= 256 + 3.2 + 0.01 = \mathbf{259.21}$
ii. $10.11 * 3.2 = 3.2(10 + 0.11) = 32 + 0.352 = \mathbf{32.352}$

Question 9

$u_1 = x^3, u_4 = x, u_7 = x^2$
$x = x^3 + 3d \; ; \rightarrow 2x = 2x^3 + 6d$ (1)
$x^2 = x^3 + 6d$ (2)

(1) - (2):
$2x - x^2 = 2x^3 - x^3 + 6d - 6d$
$x^3 + x^2 - 2x = 0$
$x(x^2 + x - 2) = 0$
$x(x + 2)(x - 1) = 0$
$\therefore x = 0 \; or -2 \; or \; 1$
But, $x \neq 0$ and $x \neq 1$
So, $x = \mathbf{-2} \; ; \rightarrow d = \frac{x - x^3}{3} = \frac{(-2) - (-2)^3}{3} = \mathbf{2}$

Question 10

The probability of getting an even number in the first roll apart from six, is 2/6 (number 2 and 4). If the first roll shows number six (probability 1/6), then the possible scenarios of getting a total of even score is if the second roll gives 2, 4 or 6. ie. (6,2), (6,4) and (6,6). Each of this has probability of 1/36. Hence, total probability is $2/6 + 3*(1/36) = 2/6 + 1/12 = \mathbf{5/12}$

Question 11

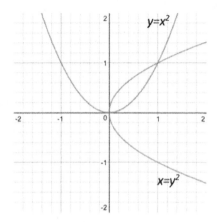

The point of intersection, as shown in the sketch is **(1,1).**

The area between the curves can be evaluated as follow:

$$A = \int_0^1 \sqrt{x} - x^2 dx = \left[\frac{2}{3}x^{\frac{3}{2}} - \frac{1}{3}x^3\right]_0^1 = \left(\frac{2}{3}(1)^{\frac{3}{2}} - \frac{1}{3}(1)^3\right) = \frac{1}{3}$$

END OF SECTION

Part B: Physics

Question 12: B

After 16000 years, element A will be ¼ of the intitial (since it has gone through 2 half-lives), and element B will be ½ of the initial. Hence, A:B = ¼ : ½ = **1:2**

Question 13: A

The voltage across resistor R_I is $V_1 = \frac{R_1}{R_1+R_2}V$ (potential divider). The power dissipated by the resistor is then $P_1 = \frac{V_1^2}{R_1} = \frac{(\frac{R_1}{R_1+R_2}V)^2}{R_1} = \frac{V^2 R_1}{(R_1+R_2)^2}$

Question 14: C

For the builder to lift the block, total moment has to be zero $\rightarrow$
$W_{concrete} l_{concrete} = F_{builder} * l_{builder}$
$(100 * 10) * 0.5 = F * 2$
$\therefore F = 250\ N$

Question 15: B

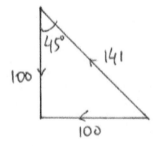

The diagram of the velocities is as above. The actual speed is then $V^2 = 141^2 - 100^2 = 9881 \approx 10000 \rightarrow V = 100\ km/hr$. **The direction is going to the left (West).**

Question 16: A

$\lambda = \frac{c}{f} = \frac{3x10^8}{1000*1000} = 300\ m$

Question 17: B

The diagram of the incident lights can be drawn as follow:

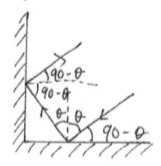

We can clearly see that the resultant beam will be 180° with respect to the incident beam.

Question 18: C

$$Q = CV = (3*10^{-9})*(10) = 3*10^{-8} \, C$$

Question 19: B

$$x = \frac{F}{k} = \frac{mg}{k} = \frac{(80)(10)}{80000} = 0.01 \, m = 10 \, mm$$

Question 20: C

From Keplar's 2nd Law, we know that $v_1 r_1 = v_2 r_2$

$$(50)(4x10^{10}) = v_2(10x10^{10})$$

$$\therefore v_2 = 20 \, km/s$$

Question 21: C

Refractive index can be thought as $n = \frac{d}{d'}$ where d' is apparent depth. The true depth d is then $= 1.33 \times 0.75 = 1 \, m$

Question 22

The observations can be written mathematically as:

A): $r = b + g$; $\rightarrow g = r - b$ (1)

B): $g^2 = 4b^2$; $\rightarrow g = 2b$ (2)

D): $\rho b^3 = 3$; $\rightarrow b^3 = \dfrac{3}{\rho}$ (3)

We then substitute (1) into (2):

$r - b = 2b$; $\rightarrow r = 3b$

$r^3 = 27b^3 = \dfrac{27(3)}{\rho}$

$m_r = \rho r^3 = 27(3) = \mathbf{81\ g}$

From (2), we know that:

$g^3 = 8b^3 = \dfrac{8(3)}{\rho}$

$m_g = \rho g^3 = 8(3) = \mathbf{24\ g}$

We know that $= \dfrac{m}{\rho}$. Hence:

$\dfrac{32}{1000} = \dfrac{24}{\rho}$

$\therefore \rho = \mathbf{750\ kg\ m^{-3}}$

Question 23

The total energy absorbed by the frying pan $E = \bar{P}At = m_s c_s \Delta T$.

The time taken to reach temperature of 70°C is:

$t = \dfrac{m_s c_s \Delta T}{\bar{P}A} = \dfrac{(2)(490)(70-20)}{10^3(0.07)} = \mathbf{700\ s}$

The final temperature of the water will be reached when both the frying pan and the water is at same temperature:

$m_s c_s (70 - T_f) = m_w c_w (T_f - 20)$

$\rightarrow (2)(490)\left(70 - T_f \right) = (4)(4200)(T_f - 20)$

$\rightarrow 17780 T_f = 404600$

$\therefore T_f = \mathbf{22.8^o C}$

Question 24

Based on the information from the question, we can sketch the system as below:

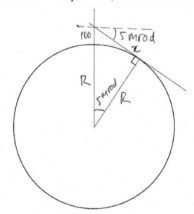

We can then deduce $\cos(5x10^{-3}) = \frac{R}{R+100}$; but we know that for small angle →
$\cos\theta \approx 1 - \frac{\theta^2}{2}$. Hence:

$1 - \frac{(5*10^{-3})^2}{2} = \frac{R}{R+100}$;

$2 - 25 * 10^{-6} = \frac{2R}{R+100}$

$2R + 200 - 25 * 10^{-6}R - 2.5 * 10^{-3} = 2R$

$R = \frac{200+0.0025}{2.5*10^{-5}} = \mathbf{8.0 * 10^6\ m}$

To find x, we know that:

$(R + 100)^2 = R^2 + x^2$

$\rightarrow R^2 + 200R + 10000 = R^2 + x^2$

$\rightarrow x^2 = 200R + 10000 = 16 * 10^8$

$\therefore \boldsymbol{x = 4 * 10^4\ m}$

Question 25

In order for the projectile to land in the rail car, the horizontal component of its velocity has to equal the rail car velocity.

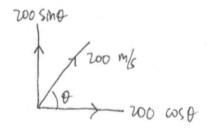

$\rightarrow 200 \cos \theta = 100$

$\rightarrow \cos \theta = \dfrac{1}{2}$

$\therefore \boldsymbol{\theta = 60^o}$

The vertical velocity of the projectile as it is fired is $200 \sin (60) = 100\sqrt{3}$. Its movement is decelerated by gravity ($a = 10$ m/s^2). When reaching the maximum altitude, its velocity become zero. Hence:

$v = u + at$

$0 = 100\sqrt{3} + (-10)t$

$\therefore \boldsymbol{t = 17.3 \, s}$

When the projectile lands in the car, the total time taken will be twice the time taken to reach maximum altitude (by "symmetry" of the projection and movement).

Hence, at that time, the rail car will be at a distance of $d = vt = (100)(2 *$ $17.3) = 3460 \, m = \boldsymbol{3.46 \, km}$

Maximum altitude can be calculated as follow:

$s = ut + \dfrac{1}{2}at^2$

With u being the initial vertical velocity of the projectile, a is gravitational acceleration and t is the time taken to reach the maximum altitude. Hence:

$\therefore s = \left(100\sqrt{3}\right)(17.3) + \dfrac{1}{2}(-10)(17.3)^2 = 1500 \, m = \boldsymbol{1.5 \, km}$

Evaluating kinetic energy:

$$KE_{rail\ car} = \frac{1}{2}mv^2 = \frac{1}{2}(200)(100)^2 = \mathbf{1\ MJ}$$

$$KE_v = \frac{1}{2}mv_v^2 = \frac{1}{2}(10)\left(100\sqrt{3}\right)^2 = \mathbf{150\ kJ}$$

$$KE_h = \frac{1}{2}mv_h^2 = \frac{1}{2}(10)(100)^2 = \mathbf{50\ kJ}$$

During the vertical movement, the vertical kinetic energy will be converted into potential energy. Hence:

$$KE_v = mgh$$
$$\therefore h = \frac{KE_v}{mg} = \frac{150000}{(10)(10)} = 1500\ m = \mathbf{1.5\ km}$$

When the projectile lands in the rail car, the velocity of the car does not change because:

i)the vertical component of the projectiles velocity cannot affect the car as they are perpendicular

ii)Horizontally, since both car and projectile are travelling with the same velocity (100 m/s), hence the final velocity of both will stay the same.

Total kinetic energy after the projectile lands will be:

$$KE_{total} = \frac{1}{2}mv^2 = \frac{1}{2}(210)(100)^2 = \mathbf{1.05\ MJ}$$

END OF PAPER

2011

Part A:

Question 1

$\cos^2\theta = 1 - \sin^2\theta$; *hence* $\rightarrow \sin\theta - 2\cos^2\theta + 1 = \sin\theta - 2(1 - \sin^2\theta) + 1 = 0$

$\rightarrow 2\sin^2\theta + \sin\theta - 1 = 0$

$\rightarrow (2\sin\theta - 1)(\sin\theta + 1) = 0$

$\rightarrow \sin\theta = \frac{1}{2}$; *or* $\sin\theta = -1$

$\therefore \theta = \frac{\pi}{6}, \frac{5\pi}{6}, \textit{or} \frac{3\pi}{2}$

Question 2

In this function, the *sin* term will vary from -1 to 1. Hence, the overall function will have a maximum of 3 + 2(1) = 5 and minimum 3 + 2(-1) = 1. When *x=0*, we can evaluate the *sin* term to be sin(-π) = 0 (ie. *y* = 3). The *sin* term will be minimum when the angle is $-\frac{\pi}{2}$ *or* $\frac{3}{2}\pi$ → *x* = 1.5 or 7.5.

Sin term will be maximum when its angle is $\frac{\pi}{2}$ *or* $\frac{5\pi}{2}$ → *x* = 4.5 or 10.5.

We can then sketch the function to be as follow:

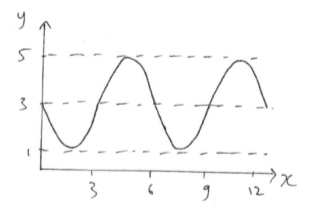

Question 3

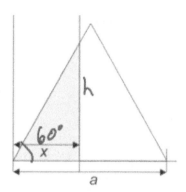

i) for $0 \leq x \leq \frac{a}{2}$

$\tan 60 = \frac{h}{x}$; $\rightarrow h = \sqrt{3}x$

$A = \frac{1}{2}xh = \frac{\sqrt{3}}{2}x^2$

ii) for $\frac{a}{2} \leq x \leq a$

when $x = \frac{a}{2}$, $h = \frac{\sqrt{3}a}{2}$

Area of whole triangle $= \frac{1}{2}a\frac{\sqrt{3}a}{2} = \frac{\sqrt{3}}{4}a^2$

$\rightarrow A = \frac{\sqrt{3}}{4}a^2 - \frac{\sqrt{3}}{2}(a-x)^2$

Question 4

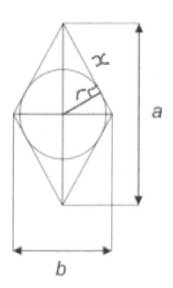

Using similar triangles principle:

$\frac{b/2}{r} = \frac{x}{a/2}$ $\rightarrow$ $r = \frac{ab}{4} * \frac{1}{x} = \frac{ab}{4} *$

$\frac{1}{\sqrt{\left(\frac{a}{2}\right)^2 + \left(\frac{b}{2}\right)^2}}$

$\rightarrow A_c = \pi r^2 = \pi \left(\frac{ab}{4}\right)^2 \frac{4}{a^2 + b^2} =$

$\frac{\pi a^2 b^2}{4(a^2+b^2)}$

$\rightarrow A_r = \frac{1}{2}ab$

$\therefore \frac{A_c}{A_r} = \frac{\pi a^2 b^2}{4(a^2+b^2)} * \frac{2}{ab} = \frac{\pi ab}{2(a^2+b^2)}$

Question 5

If $2^x = 10$; $then \rightarrow x \log 2 = \log 10$

$x = \dfrac{\log 10}{\log 2} = \dfrac{\log 10}{\log 10 - \log 5} = \dfrac{\log 10}{\log 10 - 0.7} = \dfrac{1}{1 - 0.7} = \mathbf{3.33}$

Question 6

$\sum_{r=1}^{6} \left(2^r + \dfrac{2r}{3}\right) = \left(\sum_{r=1}^{6} 2^r\right) + \dfrac{2}{3}\left(\sum_{r=1}^{6} r\right) = (2^1 + 2^2 + 2^3 + 2^4 + 2^5 + 2^6) +$

$\dfrac{2}{3}(1 + 2 + 3 + 4 + 5 + 6)$

$= 2 + 4 + 8 + 16 + 32 + 64 + \dfrac{2}{3}(21) = 126 + 14 = \mathbf{140}$

Question 7

We need to divide the original equation with the known factor $(x^2 - x - 6)$. We then get:

$0 = x^4 + 4x^3 - 17x^2 - 24x + 36 = (x^2 - x - 6)(x^2 + 5x - 6)$

$0 = (x - 3)(x + 2)(x + 6)(x - 1)$

$\therefore x = \mathbf{-6, -2, 1, 3}$

Question 8

(i) $\qquad \int \dfrac{x+2}{(x+1)(x-1)} dx$

First, we need to separate the function into two $\rightarrow \dfrac{x+2}{(x+1)(x-1)} = \dfrac{a}{(x+1)} +$

$\dfrac{b}{(x-1)} = \dfrac{ax - a + bx + b}{(x+1)(x-1)}$

Matching the variables:

$x\ term \rightarrow a + b = 1 \qquad\qquad (1)$

$constant \rightarrow b - a = 2 \qquad\qquad (2)$

We then get $a = -\dfrac{1}{2}$ and $b = \dfrac{3}{2}$

Hence $\rightarrow \int \dfrac{x+2}{(x+1)(x-1)} dx = \int \dfrac{-\frac{1}{2}}{(x+1)} + \dfrac{\frac{3}{2}}{(x-1)} dx = -\dfrac{1}{2}\int \dfrac{1}{x+1} dx + \dfrac{3}{2}\int \dfrac{1}{x-1} dx$

$\rightarrow \dfrac{3}{2}\ln(x - 1) - \dfrac{1}{2}\ln(x + 1) + \ln C$

$\therefore \mathbf{\ln\left(\dfrac{C(x-1)^{\frac{3}{2}}}{x+1}\right)}$

(ii) $\qquad \int_0^1 \dfrac{1}{\sqrt{x+1}} dx = \int_0^1 (x + 1)^{-\frac{1}{2}} dx = 2[(x + 1)^{\frac{1}{2}}]_0^1 = \mathbf{2(\sqrt{2} - 1)}$

Question 9

Let's call the difference between y_1 and y_2 "delta" (Δ). Hence:

$\Delta = y_1 - y_2 = x^3 - 3x^2 + 2x + 3 - x^2 + 3x + 4 = x^3 - 4x^2 + 5x + 7$

$\frac{d\Delta}{dx} = 3x^2 - 8x + 5 = 0$

$\rightarrow (3x - 5)(x - 1) = 0$

$\therefore x = \frac{5}{3} \, or \, 1$

$\frac{d^2\Delta}{dx^2} = 6x - 8$ when $x = 1 \rightarrow \frac{d^2\Delta}{dx^2} = 6(1) - 8 = -2$ that is < 0. $\rightarrow$ **maximum point**

When $x = 5/3 \rightarrow \frac{d^2\Delta}{dx^2} = 6\left(\frac{5}{3}\right) - 8 = 2$ that is >0. $\rightarrow$ **minimum point**

Question 10

First, we know that $y = \frac{t}{2x}$; $\rightarrow s = x^2 + \left(\frac{t}{2x}\right)^2$

$0 = 4x^4 - 4sx^2 + t^2$

$x^2 = \frac{4s \pm \sqrt{16s^2 - 16t^2}}{8} = \frac{s \pm \sqrt{s^2 - t^2}}{2}$

$\therefore x = \pm \sqrt{\frac{s \pm (s^2 - t^2)^{\frac{1}{2}}}{2}}$

And $y = \frac{t}{2x} = \pm \frac{t}{2} \sqrt{\frac{2}{s \pm (s^2 - t^2)^{\frac{1}{2}}}}$

Question 11

Maximum score is 35 → this will be achieved when both dice show 6 → $6A + 6B + C = 35$ (1)

Minimum score is 0 → this is when both dice show 1 → $A + B + C = 0$; $A + B = -C$ (2)

From (1) → $6(A + B) = 35 - C$

Substitute (2) into (1):

$6(-C) = 35 - C$; → $C = -7$

So $A + B = 7$.   (3)

We know that S covers all integers from 0 to 35 with an equal probability of each score. Meaning that when $d_1 = 1$ and $d_2 = 2$ (the 2nd smallest score, ie. 1):

$A + 2B - 7 = 1$

$A + 2B = 8$ (4)

Solve for (3) and (4), we get that $B = 1$ and $A = 6$.

However, when $d_1 = 2$ and $d_2 = 1$, the value of S will be the same and values for B and A will be swapped.

Hence:

$\therefore A = 6 ; B = 1$

or $A = 1 ; B = 6$

END OF SECTION

Part B:

Question 12: C

The velocity of the wave $v = \frac{s}{t} = \frac{45}{25} = \frac{9}{5} ms^{-1}$

$\lambda = \frac{v}{f} = vT = \left(\frac{9}{5}\right)(2) = \mathbf{3.6\ m}$

Question 13: A

When $Ts = 26°C$, then $\alpha(Ts - T) = (15)(26 - 6) = 300W$. Total power consumption will be $\frac{100}{30} * 300 = 1000W = \mathbf{1kW}$

Question 14: B

A lunar eclipse can only occur when then moon's phase is full moon.

Question 15: D

Power radiated will be proportional to distance^2. Hence $\frac{P_d}{P_n} = \frac{r_d^2}{r_n^2} = \left(\frac{20}{10}\right)^2 = \mathbf{4}$

Question 16: A

Kepler's 3rd law $\rightarrow T^2 \propto r^3$. Hence:

$\frac{T_1^2}{r_1^3} = \frac{T_2^2}{r_2^3}$; $\rightarrow r_2 = \left(\frac{T_2}{T_1}\right)^{\frac{2}{3}} r_1 = \left(\frac{24}{3}\right)^{\frac{2}{3}} (0.4) = \mathbf{1.6\ A.U.}$

Question 17: D

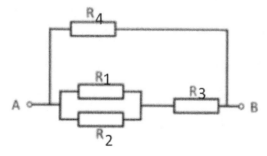

$R_{1-2} = \frac{R}{2}$

$R_{1-3} = \frac{R}{2} + R = \frac{3R}{2}$

$\therefore R_{1-4} = \frac{1}{\frac{1}{R} + \frac{2}{3R}} = \frac{1}{\frac{5}{3R}} = \frac{3R}{5}$

Question 18: C

$$\frac{V_1}{V_2} = \frac{N_1}{N_2} ; \rightarrow N_2 = \frac{V_2}{V_1} N_1 = \frac{120}{240}(50) = \mathbf{25}$$

Question 19: B

Power $= IV = Fv$. Hence: $v = \frac{IV}{F} = \frac{IV}{mg} = \frac{(1)(6)}{(0.1)(10)} = \mathbf{6ms^{-1}}$

Question 20: D

The component of the weight parallel with the slope $F = mg \sin 30 = ma$
$a = g \sin 30 = (10)(0.5) = \mathbf{5ms^{-2}}$

Question 21: C

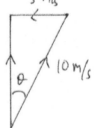

$\sin \theta = \frac{5}{10} = \frac{1}{2} ; \rightarrow \theta = 30^o$ and she needs to steer in **east** direction to overcome the river current.

Question 22

For the first case, the spring potential energy will be converted into kinetic energy. Hence:

$$\frac{1}{2}k(\Delta x)^2 = \frac{1}{2}mv^2 ; \rightarrow v = \sqrt{\left(\frac{k}{m}\right)} \Delta x = \sqrt{\left(\frac{k}{m}\right)}(x - l)$$

In the second case, the spring potential energy is now converted into kinetic energy **and** gravitational potential energy of height *(x-l)*. Hence:

$$\frac{1}{2}k(x - l)^2 = \frac{1}{2}mv^2 + mg(x - l)$$
$$\rightarrow v^2 = \frac{k}{m}(x - l)^2 - 2g(x - l)$$
$$\therefore v = \sqrt{\frac{k}{m}(x - l)^2 - 2g(x - l)}$$

When thrown vertically, all kinetic energy eventually converts into gravitational potential energy. Hence:

$$\frac{1}{2}mv^2 = mgh ; \rightarrow h = \frac{k}{2gm}(x - l)^2 - (x - l)$$

Question 23

To fly undeflected, eE has to equal evB. Hence we can first find the speed at which the electron has to fly:

$$eE = evB \; ; \rightarrow v = \frac{E}{B} = \frac{1000}{1*10^{-3}} = 10^8 \; ms^{-1}$$

This speed (ie. kinetic energy) is provided by the accelerating voltage through electrical potential energy. Hence:

$$\frac{1}{2}mv^2 = eV \; ; \rightarrow V = \frac{\frac{1}{2}(10^{-30})(10^8)^2}{1.6*10^{-19}} = \mathbf{3.1 * 10^4 \; V}$$

Question 24

A sheet of aluminium will stop beta-radiation. Hence from observation (i), we know that $\beta = 50$.

When the source is taken away completely, the detected radiation is background radiation (neither alpha nor beta nor gamma). Gamma is then 100-50-10 = 40

When the source is placed 1 cm from the detector, we will also detect alpha-radiation. Hence we know that $\alpha = 300$.

Alpha:beta:gamma is then 300:50:40 = **30:5:4**

Question 25

In this question, please note that there is a mistake in the question script. Statement C should have said *"The base area (i.e. width times length) of a large box is 9 times larger than the base area of the small box."*

Let $\frac{height}{width} = x$ and $\frac{length}{width} = y$

Then, statements A-E can be expressed mathematically as follow:

$$A \rightarrow 8w_s h_s l_s = w_m h_m l_m \qquad (1)$$
$$B \rightarrow l_s = h_m \qquad (2)$$
$$C \rightarrow 9w_s l_s = w_l l_l \qquad (3)$$
$$D \rightarrow l_s + l_m + l_l = 2.4 \qquad (4)$$
$$E \rightarrow w_m = 2h_s \qquad (5)$$

Substitute x and y into (1):
$$8w_s(xw_s)(yw_s) = w_m(xw_m)(yw_m)$$
$$\rightarrow 8w_s^3 = w_m^3$$
$$\therefore 2w_s = w_m \; ; \rightarrow 2h_s = h_m \, , 2l_s = l_m$$

Substitute x and y into (3):
$$9w_s(yw_s) = w_l(yw_l)$$
$$\rightarrow 9w_s^2 = w_l^2$$
$$\therefore 3w_s = w_l \; ; \rightarrow 3h_s = h_l \, , 3l_s = l_l$$

Equation (4) then becomes:
$$l_s + 2l_s + 3l_s = 2.4 \; ;$$
$$\therefore l_s = 0.4 \, m \; ; \; l_m = 0.8 \, m; \; l_l = 1.2 \, m$$

From equation (5), we know that:
$$w_m = 2h_s = 2w_s \; ; \rightarrow \frac{w}{h} = 1$$

From equation (2):
$$l_s = h_m = 2h_s \; ; \rightarrow \frac{h}{l} = \frac{w}{l} = \frac{1}{2}$$

Question 26

The elastic potential energy of the stretched string $E = \frac{1}{2}fx = \frac{1}{2}(120)(0.6) = 36\,J$

This will be equal to kinetic energy of the arrow $\frac{1}{2}mv^2$. Hence:

$$36 = \frac{1}{2}mv^2 \; ; \to v = \sqrt{\frac{36(2)}{0.02}} = \textbf{60 } \boldsymbol{ms^{-1}}$$

The actual kinetic energy is $\frac{25}{36} * 36 = 25\,J = \frac{1}{2}mv^2 \; ; \to v = \boldsymbol{50ms^{-1}}$

With constant speed, time taken is $t = \frac{distance}{speed} = \frac{50}{50} = \textbf{1 } \boldsymbol{s}$

The vertical distance travelled due to gravity over 1 s can be calculated as follow:
$s = ut + \frac{1}{2}gt^2 = (0)(1) + \frac{1}{2}(10)(1)^2 = \textbf{5 } \boldsymbol{m}.$

Hence the arrow has to be aimed 5 m above the centre to overcome this gravitational drop.

The horizontal speed of the arrow v_h is 50 ms^{-1}. The vertical speed is $v_v = u + gt = (0) + (10)(1) = 10\,ms^{-1}$
Total velocity is then $V^2 = 50^2 + 10^2 = 2600$

We know that $\frac{1}{2}mV^2 = Fx \; ; \to F = \frac{\frac{1}{2}mV^2}{x} = \frac{\frac{1}{2}(0.02)(2600)}{0.005} = \textbf{5200 } \boldsymbol{N}$

Using conservation of momentum,

$momentum\ of\ arrow + momentum\ of\ target\ (before\ collision)$
$= momentum\ of\ arrow + target\ (after\ collision)$

$\to (0.02)\left(\sqrt{2600}\right) + 0 = (5.02)v \; ; \to v = \frac{0.02(\sqrt{2600})}{5.02} = \boldsymbol{0.2\ ms^{-1}}$

END OF PAPER

2012

Part A:

Question 1

We can sketch the functions as follow:

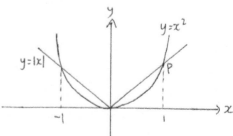

At P, $x = x^2$

$\rightarrow 0 = x^2 - x = x(x - 1)$

$\rightarrow x = 0 \; or \; x = 1$

Area can be calculated by integration, noting that the area is symmetrical about the y-axis:

$A = 2 \int_0^1 (x - x^2) dx = 2[\frac{1}{2}x^2 - \frac{1}{3}x^3]_0^1 = 2\left(\frac{1}{2} - \frac{1}{3}\right) = \frac{1}{3}$

Question 2

i) this should be straightforward, and you should remember binomial expansion formula.

$(4 + x)^4 = 4^4 + 4(4)^3 x + 6(4)^2 x^2 + 4(4)x^3 + x^4 = \mathbf{256 + 256x + 96x^2 + 16x^3 + x^4}$

ii) In this case, $x = 0.2$. Hence $\rightarrow$ $(4.2)^4 = 256 + 256(0.2) + 96(0.2)^2 + 16(0.2)^3 + (0.2)^4 = 256 + 51.2 + 3.84 + 0.128 + 0.0016 = \mathbf{311.17}$

Question 3

$\sum_{r=1}^{8}(2 + 4^r) = \sum_{r=1}^{8} 2 + \sum_{r=1}^{8} 4^r = (8 * 2) + (4 + 4^2 + 4^3 + \cdots + 4^8)$

The 2nd term of the summation is a geometric progression with first term = 4 and common ratio 4, hence:

$= (8 * 2) + (4 + 4^2 + 4^3 + \cdots + 4^8) = 16 + \dfrac{4(4^8 - 1)}{4 - 1} = 16 + \dfrac{4(65535)}{3}$

$= \mathbf{87396}$

Question 4

The rough sketch of the image can be seen below:

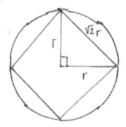

The side of the square is $\sqrt{2}r$. Hence, the area of the shaded region:

$$A_{shaded} = A_{circle} - A_{square}$$
$$A_{shaded} = \pi r^2 - \left(\sqrt{2}r\right)^2$$
$$\therefore \boldsymbol{A_{shaded} = (\pi - 2)r^2}$$

Question 5

To prove that $x = 1$ is a solution to the function, we substitute $x = 1$ into the function:

$$(1)^3 - 6(1)^2 - 9(1) + 14 = 1 - 6 - 9 + 14 = 0$$

This means that $(x - 1)$ is a factor of the function. The function can then be factorised into:

$$x^3 - 6x^2 - 9x + 14 = (x - 1)(x^2 - 5x - 14) = 0$$
$$\rightarrow (x - 1)(x - 7)(x + 2) = 0$$
$$\therefore \boldsymbol{x = 1 \; or \; 7 \; or \; -2}$$

Question 6

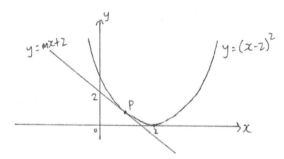

At P, $y = mx + 2 = (x - 2)^2$
$mx + 2 = x^2 - 4x + 4$
$0 = x^2 - (4 + m)x + 2$
Also, at P the gradient of the curve equals the gradient of the line:
$\frac{dy}{dx} = 2(x - 2) = m$
$m = 2x - 4$

Substitute m into the first equation above:
$0 = x^2 - (4 + 2x - 4)x + 2$
$0 = x^2 - 2x^2 + 2$
$x^2 - 2 = 0 ; \rightarrow x = \pm\sqrt{2}$

For $x > 0$, $m = 2\sqrt{2} - 4$
$\therefore y = \left(2\sqrt{2} - 4\right)x + 2$

Question 7

First, we know that $\log_2 16 = 4$ and $\log_{10} \sqrt{0.01} = \log_{10} 0.1 = -1$. Hence:
$5 = \log_2 16 + \log_{10} \sqrt{0.01} + \log_3 x$
$5 = 4 - 1 + \log_3 x ;$
$\log_3 x = 2 ;$
$\therefore x = 9$

Question 8

To obtain a total score of 7, then possible scenarios are (dice 1(1-6), dice 2(1-3)):
$(6,1) = 1/6 * 1/3 = 1/18$
$(5,2) = 1/6* 1/3 = 1/18$; and
$(4,3) = 1/6 * 1/3 = 1/18$

Total probability $= 3/18 = 1/6$

Question 9

$\cos^2 \theta + \sin \theta = 0$;

$(1 - \sin^2 \theta) + \sin \theta = 0$

$\sin^2 \theta - \sin \theta - 1 = 0$;

This is a quadratic function of $\sin \theta$, hence it can be solved using the abc formula:

$\sin \theta = \frac{1 \pm \sqrt{1 - 4(1)(-1)}}{2} = \frac{1 \pm \sqrt{5}}{2}$; but $\sin \theta \leq 1$

$\therefore \sin \theta = \frac{1 - \sqrt{5}}{2}$

Question 10

From statement (c), we know that at $x = 4$ the tangent of the curve $= 0$ (ie. maximum/minimum/inflection point). However, we know that it cannot be i) a maximum point since y cannot be negative (statement (a)); nor ii) an inflection point from statement (d). So, at $x = 4$, it has to be a minimum point. Furthermore, the inflection points are at $x = 2$ and $x = 6$ (second derivative equals zero). Hence, an example of the function is:

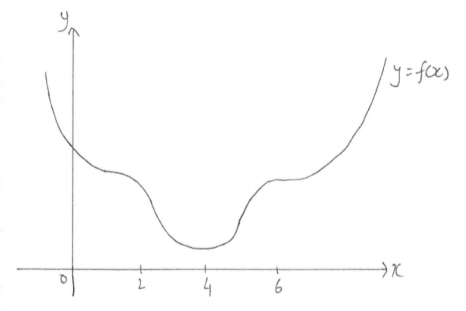

Please note also that the function has to be continuous (no sudden change in slope), ie. requirement (b).

Question 11

We can split this inequality into two:

$$-1 < -\frac{1}{x} + 2x \quad (1); \text{ and}$$
$$-\frac{1}{x} + 2x < 1 \quad (2)$$

(1):
$$-1 < -\frac{1}{x} + 2x \; ;$$
$$-x^2 < -x + 2x^3 \; ;$$
$$0 < 2x^3 + x^2 - x;$$
$$0 < x(2x - 1)(x + 1);$$

We then make a number line as follow:

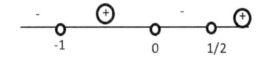

(2):
$$-\frac{1}{x} + 2x < 1$$
$$-x + 2x^3 < x^2$$
$$2x^3 - x^2 - x < 0$$
$$x(2x + 1)(x - 1) < 0$$

Combining the two number lines, we get:

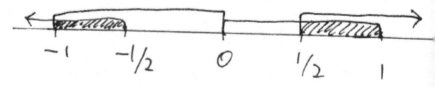

$$\therefore \left\{-1 < x < -\frac{1}{2}\right\} \cup \left\{\frac{1}{2} < x < 1\right\}$$

Question 12

$y = x^{-2} - x^{-1} - 1$

$\frac{dy}{dx} = -2x^{-3} + x^{-2} = 0$;

$\rightarrow -2 + x = 0$;

$x = 2 \rightarrow turning\ point$

- At $x = 2$, $y = -5/4$
- As $x \rightarrow \infty$, $y \rightarrow -1$; $when\ x \rightarrow -\infty$, $y \rightarrow -1$
- $x = 0$ is asymptote

The sketch will then look like as follow:

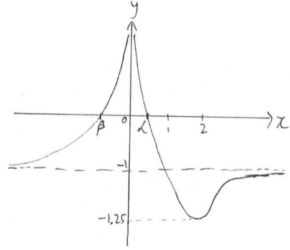

To find α, β ; $y = 0$.

$\rightarrow 1 - x - x^2 = 0$

$\alpha, \beta = \frac{-1 \pm \sqrt{1+4}}{2} = \frac{-1 \pm \sqrt{5}}{2}$

END OF SECTION

Part B:

Question 13: C

The ratio between mass : volume of both the actual size and its model will be constant, ie. mass $\propto$ length^3 since both are made out of iron (same density). Hence,

the length of the model will be $l^3 = \frac{1}{6.5*10^4} * 10^3$; $l = \left(\frac{1}{6.5*10^4}\right)^{\frac{1}{3}} * 10 \approx \boldsymbol{25\ cm}$

Question 14: D

For ideal gas, $P = \frac{nRT}{V}$; $n = \frac{m}{Mr}$.

Hence: $P = \frac{mRT}{Mr.V} = \frac{(88)(8.3)(273+27)}{(44)(0.02)} = 249,000\ Pa = \boldsymbol{249\ kPa}$

Question 15: B

In this case, $P = IV = Fv$.

F here will be the forward driving force of the car, which is equal to the air resistance since the car is moving with a constant speed (zero acceleration). Hence

$F = \frac{IV}{v} = \frac{(100)(160)}{\left(\frac{36}{3.6}\right)} = \boldsymbol{1600\ N}$

Question 16: C

The ones painted will be the those originally on the surface of the big cube. As the cube is cut into 125 pieces, each side has to be divided by $\sqrt[3]{125} = 5$. So the number of little cubes that will be painted can be determined as follow:

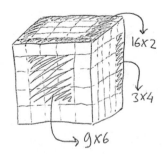

No. of painted cubes are 9x6 + 16x2 +3x4 = 98. Hence the unpainted ones are 125 − 98 = **27**

Question 17: A

The slider goes not get to T as the curvature of the track introduces some centripetal forces that need to be spent by the slider, reducing the kinetic energy, hence the speed available for the slider to get to T.

Question 18

Voltmeter has to be placed parallel to the resistor, whereas ammeter has to be in series. Hence, the complete circuit will look as follow:

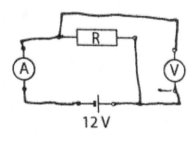

The current can be calculated as $I = \frac{V}{R} =$
$\frac{12}{2000} = 6\ mA$

When the switch is moved to e, the capacitor is charging. Hence, current variation over time will look like this:

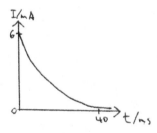

The time constant $\tau = RC = 2 * 4x10^{-3} = 8\ ms$.

T can be estimated as $5\tau = 40\ ms$

When the switch is moved to d, discharging of the capacitor happens. Hence the current will look like this:

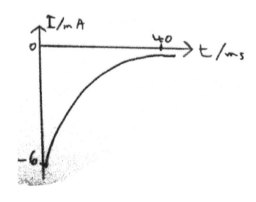

Question 19

The function of the screen is to reflect the soundwave coming from the loudspeaker. Between two successive maxima, the path difference of the (reflected) waves coming from the screen (and received by the microphone) has to be λ. In order to achieve that, thus the distance between the two positions must be $\frac{\lambda}{2}$, as the wave will travel forth and back of that distance, i.e. incident and reflected.

Maxima (or minima) is achieved when two waves constructively (or destructively) interfere. In this case, the wave sources are the loudspeaker and the screen. To observe maxima and minima, one needs to alter one of the sources. Hence, **moving the loudspeaker (a) will achieve this, however moving the microphone (b) will not.**

Question 20

$$r(t) = \frac{N_{235}(t)}{N_{238}(t)} = \frac{N_{235}^{o} \, e^{-\lambda_{235}t}}{N_{238}^{o} \, e^{-\lambda_{238}t}}$$

For $t = 10^9$ years,

$$r(t) = 0.0072 = \frac{N_{235}(t)}{N_{238}(t)} = \frac{N_{235}^{o} \, e^{-\lambda_{235}(10^9)}}{N_{238}^{o} \, e^{-\lambda_{238}(10^9)}}$$

We know that $\lambda = \frac{\ln(2)}{t_{1/2}}$; and says $\frac{N_{235}^{o}}{N_{238}^{o}} = r_o$, hence:

$$0.0072 = r_o \frac{e^{-\frac{\ln(2)}{7 \times 10^8}(10^9)}}{e^{-\frac{\ln(2)}{4.5 \times 10^9}(10^9)}} = r_o \frac{e^{-\frac{0.7}{0.7}}}{e^{-\frac{0.7}{4.5}}} = \frac{r_o(2.7)^{-1}}{1 - \frac{0.7}{4.5}}$$

$$Hence, r_o = 0.0072 * \frac{38}{45} * 2.7 \approx \mathbf{0.016}$$

Question 21

We know that gravitational potential force = centripetal force. Hence:

$$\frac{GMm}{R^2} = \frac{mv^2}{R} \rightarrow v = \sqrt{\frac{GM}{R}}$$

Where G = gravitational constant, M is the mass of earth, and R is the orbit radius.

The trajectory of the double mass will look as follow:

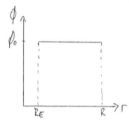

The azimuthal angle is independent on distance r.

KE against r will look like this:

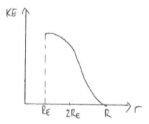

As $KE = \frac{1}{2} mv^2$, and $GPE = -GMm/r \rightarrow KE$ is proportional to $1/r$

Temperature against r :

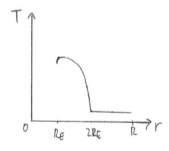

As there's no heating outside the earth's atmosphere, causing constant temperature.

Question 22
A1
Throughout the motion, kinetic energy equals work done (as there is no friction).
At $x = 0$,
$$\frac{1}{2}mv^2 = Fx$$
$$v = \sqrt{\frac{2Fx}{m}} = \sqrt{\frac{2*10*10}{1}} = \sqrt{200} \approx 14\ ms^{-1}$$

A2
Kinetic energy = work done = Fx ; hence, KE will be the area under the F-x plot, ie. integration of F-x plot. It will then look as follow:

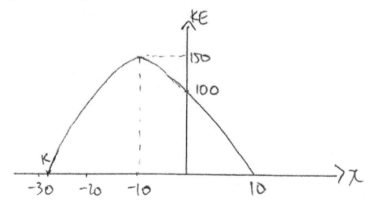

Max. kinetic energy is achieved when F (ie. the gradient) is zero ($x = -10$ m). At this point, area under the F-x curve from $x = -10$ m to $x = 10$ m $= 10 * 10 + \frac{1}{2} * 10 * 10 = 150\ J$

Point K is x when kinetic energy is zero, meaning that the area above the x-axis equals the area below it, ie. 150 J.

Area above x-axis $= 150 = \frac{1}{2}(-10 - K)^2$

$\therefore K = -27.3\ m$

A3

As $F=ma$ and $m = 1$ kg, then $F=a$.

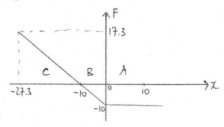

At A: $s = -10$; $u = 0$; $v = -14$; $a = -10$; $t = ?$
$v = u + at$;
$-14 = -10t$
$t = 1.4\ s$

B: $u = 14$; $v = 17.3$
$\rightarrow \frac{1}{2} \times 10 \times t = 17.3 - 14$;
$t = 0.64\ s$

C: $\frac{1}{2} \times 17.3 \times t = 17.3 - 0$;
$t = 2$ s

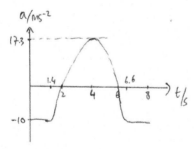

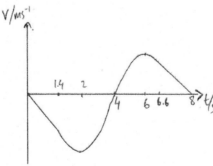

B1

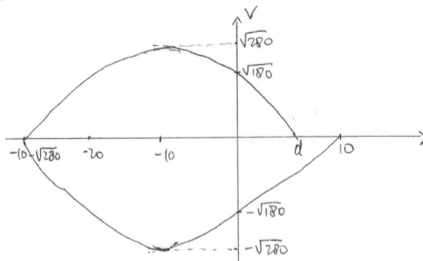

$$Max.KE = (9 \; x \; 10) + \left(10 \; x \; 10 \; x \frac{1}{2}\right) = 140 \, J$$

$\rightarrow Max. v = \sqrt{280}$

$$d = \frac{1}{2} \; x \frac{\left(\sqrt{180}\right)^2}{11} = 8.2 \, m \; (2 \, sf)$$

B2

Distance $= 10 + d = 10 + 8.2 = \textbf{18.2 m}$

END OF PAPER

2013

Part A:

Question 1

This is a geometric progression with common ratio $r = -1/3$ and first term $a = 2/3$.

$$\therefore S_\infty = \frac{a}{1-r} = \frac{\frac{2}{3}}{1-\left(-\frac{1}{3}\right)} = \frac{1}{2}$$

Question 2

Substitute y into u:

$$u = \left(\sqrt{xv} - \sqrt{x}\right)^2 = xv - 2x\sqrt{v} + x\ ;$$
$$u = x\left(v - 2\sqrt{v} + 1\right);$$
$$\therefore x = \frac{u}{v - 2\sqrt{v} + 1}$$

Question 3

The events are independent, so:

a) $P(female \cap red\ hair) = P(female) * P(red\ hair) = \frac{30}{50} * \frac{8}{50} = \frac{240}{2500} = \frac{12}{125}$

b) $P(male \cap other\ colour) = P(male) * P(other\ colour) = \frac{20}{50} *$ $\left(1 - \frac{11}{50}\right) = \frac{780}{2500} = \frac{39}{125}$

Question 4

a) Put $x = 1$ into $f(x)$ → $(1)^3 - (1)^2 - 4(1) + 4 = 0$.

Hence, $(x-1)$ is a factor. We can then factorise $f(x)$ into:

$$f(x) = x^3 - x^2 - 4x + 4 = (x-1)(x^2 - 4) = (x-1)(x+2)(x-2)$$
$$\therefore x = 1\ or -2\ or\ 2$$

b) Area under the curve $= \int_{-2}^{1} x^3 - x^2 - 4x + 4\ dx = [\frac{1}{4}x^4 - \frac{1}{3}x^3 - 2x^2 +$ $4x]_{-2}^{1} = \left(\frac{1}{4} - \frac{1}{3} - 2 + 4\right) - \left(\frac{1}{4}(-2)^4 - \frac{1}{3}(-2)^3 - 2(-2)^2 + 4(-2)\right) = \frac{45}{4}$

Question 5

Equate both expressions of x:

$\log_{10} 100 + \log_5 \sqrt{25} - \log_3 y^2 = 2(\log_2 8 - 9\log_{10}\sqrt{10} + 2\log_3 y)$

$\rightarrow 2 + 1 - 2\log_3 y = 2(3 - 9\left(\frac{1}{2}\right) + 2\log_3 y)$

$\rightarrow 3 - 2\log_3 y = -3 + 4\log_3 y$

$\rightarrow 6 = 6\log_3 y$

$\therefore y = 3$

$x = \log_{10} 100 + \log_5 \sqrt{25} - \log_3 y^2 = 2 + 1 - \log_3(3)^2 = 2 + 1 - 2$

$\therefore x = 1$

Question 6

For *C1*: $x^2 + 4x + y^2 - 2y = -1$; $\rightarrow$ *the centre of the circle is* $(-2,1)$

For *C2*: $x^2 - 4x + y^2 - 6y = 3$; $\rightarrow$ *the centre of the circle is* $(2,3)$

Hence, the equation of the line:

$(y - y_1) = \left(\frac{y_2 - y_1}{x_2 - x_1}\right)(x - x_1)$;

$\rightarrow (y - 1) = \left(\frac{3-1}{2-(-2)}\right)(x - (-2))$

$\therefore y = \frac{1}{2}x + 2$

Question 7

$(3.12)^5 = (3 + 0.12)^5 = [3(1 + 0.04)]^5 = 3^5(1 + 0.04)^5$

$= 3^5[1 + 5(0.04) + 10(0.04)^2 + 10(0.04)^3 + \cdots)$

$= 3^5[1 + 0.20 + 0.016 + 0.00064 + \cdots)$

$3^5 * 0.00064 = 0.16 \rightarrow$ *this is one decimal place, hence the term required*

$\therefore$ *so 4 terms in total*

Question 8

From the two inequalities, we can deduce the following curves:

$$y > \frac{1}{x} \; ; y < \frac{2}{x} \; ; y > \frac{1}{2}x \; ; y < 2x$$

Hence, the regions will look as follow:

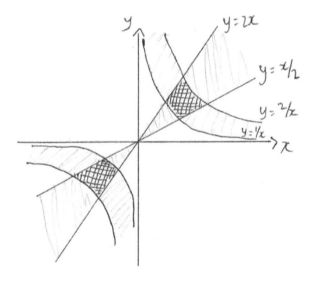

Question 9

a) This sketch should be straightforward, and you should be familiar with it. $y = \exp(-x)$ will look as follow:

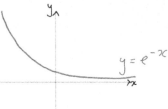

b) $y = 3\{\exp[-2(x-1)] - 2\exp[-(x-1)]\}$; first we need to find what happens to y as x goes to +/- infinity:

$as\ x \to +\infty, y \to 0\ (from - ve)$

$as\ x \to -\infty, y \to \infty$

Then, when $y = 0$,

$3\{\exp[-2(x-1)] - 2\exp[-(x-1)]\} = 0$;

$e^2 e^{-2x} - 2e.e^{-x} = 0$;

$e^{-x}(e^2 e^{-x} - 2e) = 0$;

$e^{-x} = 0\ or\ e^{-x} = \dfrac{2}{e}$

→ $e^{-x} = 0$ has no solutions, whereas $e^{-x} = \dfrac{2}{e}$ means that the curve cuts the x-axis once on positive x.

When $x = 0$,

$y = 3(e^2 - 2e)\ which\ is\ > 0$;

→ this means the curve cuts the y-axis once on positive y

Next, we need to find any turning points:

$\dfrac{dy}{dx} = 3\left[-2e^{-2(x-1)} + 2e^{-(x-1)}\right] = 0$;

$-e^2.e^{-2x} + e.e^{-x} = 0$;

$e^{-x}(e^2 e^{-x} - e) = 0$;

$e^{-x} = 0\ or\ e^{-x} = \dfrac{1}{e} > 0$;

→ this means there is one turning point on positive x, and it happens at x bigger than where the curve meets the x-axis.

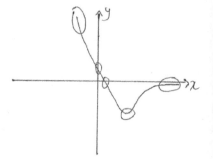

Question 10

We can introduce a few variables into the figure as follow:

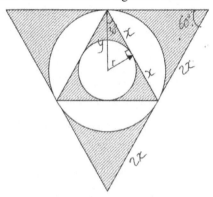

Looking at the right-angle triangle inside the smaller triangle:

$\tan(30) = \frac{r}{x}$;

$\rightarrow x = \frac{r}{\frac{1}{\sqrt{3}}} = \sqrt{3}r$;

$\sin(30) = \frac{r}{y}$;

$\rightarrow y = \frac{r}{\frac{1}{2}} = 2r$

Area of small circle $= \pi r^2$

Area of big circle $= \pi(2r)^2 = 4\pi r^2$

Area of small triangle $= \frac{1}{2}(2\sqrt{3}r)(2\sqrt{3}r)\sin(60) = 3\sqrt{3}r^2$

Area of big triangle $= \frac{1}{2}(4\sqrt{3}r)(4\sqrt{3}r)\sin(60) = 12\sqrt{3}r^2$

Then, shaded area $= (12\sqrt{3} - 4\pi + 3\sqrt{3} - \pi)r^2 = (15\sqrt{3} - 5\pi)r^2 = \mathbf{5r^2(3\sqrt{3} - \pi)}$

END OF SECTION

Part B:

Question 11: B

Powers are constant across the transformer, hence $I_p V_p = I_s V_s$. Hence, $V_s = \frac{2.4*100}{4.8} = 50\ V$.

$\frac{N_p}{N_s} = \frac{V_p}{V_s}$; $N_s = \frac{V_s}{V_p} N_p = \frac{50}{100} * 100 = 50\ turns$

Question 12: C

We know that A will be halved every 3 days, whereas B will be halved every 6 days. Says initially we have 80 A and 40 B.

After 6 days, A will be ¼ of initial amount = 20, whereas B will be ½ of its initial amount = 20.

After 12 days, A will be further quartered = 5, whereas B will be further halved = 10. The ratio of A:B is now 1:2.

Question 13: A

We know that when two resistors are in parallel, their total resistance become $R_T = \frac{R_1 R_2}{R_1 + R_2}$. Hence:

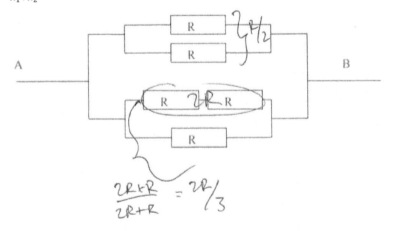

Then, $R_T = \frac{R_1 R_2}{R_1 + R_2} = \frac{(\frac{R}{2})(\frac{2R}{3})}{\frac{R}{2} + \frac{2R}{3}} = \frac{2R}{7}$

Question 14: B

From Kepler's Law, we know that $T^2 \propto r^3$

$$\rightarrow \frac{T_1^2}{T_2^2} = \frac{r_1^3}{r_2^3} = \left(\frac{r_1}{0.5r_1}\right)^3 = 8;$$

$$T_2^2 = \frac{T_1^2}{8} = \frac{(24)^2}{8} = 72 \; ;$$

$$\therefore T_2 = \sqrt{72} \approx 8.5 \; hrs$$

Question 15: C

We know that power is inversely proportional to the square of distance; ie. $P \propto \frac{1}{r^2}$

$$P_1 r_1^2 = P_2 r_2^2;$$

$$r_2^2 = \frac{P_1 r_1^2}{P_2} = \frac{(20)(100)^2}{(0.001)} = 2x10^8;$$

$$\therefore r_2 = \sqrt{2x10^8} = 10,000\sqrt{2} \; m = 10\sqrt{2} \; km$$

Question 16

Weight of the car = mg = 10,000 N. On each tyre, there will be W/4 = 2,500 N.

Pressure = Force/Area ;   Area = Force/Pressure = 2,500 / (200,000) = **0.0125 m²**

Question 17

We can sketch the forces acting on the masses as follow:

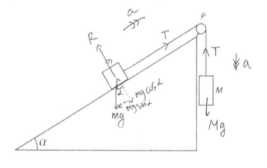

For M:

$Mg - T = Ma$

For m:

$T - mgsin(\alpha) = ma$;

$T = mgsin(\alpha) + ma$

Equate T:

$mgsin(\alpha) + ma = Mg - Ma$

;

$a(M + m) = Mg - mgsin(\alpha);$

$$\therefore a = \frac{g(M - msin(\alpha))}{(M+m)}$$

$$T = Mg - Ma = M\left(g - \frac{g(M-msin(\alpha))}{(M+m)}\right) = \frac{mgM(1-sin(\alpha))}{M+m}$$

Question 18

From conservation of momentum, we can calculate the velocity of the ball+projectile after collision:

$$m * v_p = (M + m) * v_{bp} \; ; \; v_{bp} = \frac{0.2 * 122}{12.2} = 2 \, ms^{-1}$$

Then, using the conservation of energy, we can calculate the gravitational potential energy gained by the system:

$$E_p = (M + m)gh = E_k = \frac{1}{2}(M + m)v_{bp}^2 \; ;$$

$$h = \frac{\frac{1}{2}v_{bp}^2}{g} = \frac{\frac{1}{2}(2)^2}{10} = 0.2 \, m$$

Question 19

The narrow slits act like coherent sources, which means they have a constant phase difference between them. Loght from those sources arrive at a particular point on the screen having travelled different distances. This is called path difference. Since the sources are coherent, the path difference at a particular point on the screen does not change.

When the path difference is an integer multiple of the wavelength, light from the two sources arrive in phase so there is constructive interference. This is where the bright fringes are. At points that the path difference is a half-integer multiple of the wavelength, light arrives in anti-phase so there is destructive interference. This gives rise to the dark fringes.

We need to remember that fringe separation is proportional to the wavelength. As green light has shorter wavelength than the red light, its fringe separation is smaller. A sketch will look as follow:

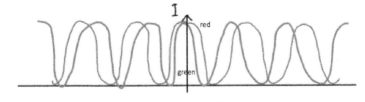

Question 20

Volume of water = volume of pot = $10*10*15 = 1500$ cm^3. Mass of water = density * volume = $1*1500 = 1500$ gr = 1.5 kg. Heat required = $mc\Delta T = (1.5)(4.2)(100-20) =$ **504 kJ.**

Reduction of boiling point is caused by the **decrease in atmospheric pressure (due to height)**. At 6000 m, the boiling point of water is $100 - (6000/300) = 80°C$. The energy required is then = $(0.1)(4.2)(80-10) =$ **29.4 kJ**

Full pot at sea level requires 504 kJ of energy and takes 15 minutes. The cup of tea requires 29.4 kJ of energy at 6000 m, and the stove only operates at 50% power. Hence, the time taken to boil the water for cup of tea is $= \frac{29.4}{504} * 15 * 2 = 1.75$ min = **105 s.**

Question 21

a. The acceleration experienced by the particle is $a = \frac{f}{m}$.

Hence: $v^2 = u^2 + 2aS = v_o^2 + \frac{2f}{m}d$;

$\therefore v = \sqrt{v_o^2 + \frac{2f}{m}d}$

b. Since force F is always perpendicular to the direction of travel (centripetal force), the motion will then be **circular** and its speed is constant.

$\rightarrow F = \frac{mv^2}{r} = \alpha v$;

$r = \frac{mv}{\alpha}$

The particle will make a path of half circle before hitting the detector, hence its vertical distance y will be $2r = \frac{2mv}{\alpha}$, where v is that found in part (a).

c. The requirement for Δy is that $\Delta y < \frac{2m(v+\Delta v)}{\alpha} - \frac{2mv}{\alpha}$. Hence:

$\Delta y < \frac{2m}{\alpha}(v + \Delta v - v)$;

$\Delta y < \frac{2m}{\alpha}(\Delta v)$;

$\therefore \Delta v > \frac{\alpha \Delta y}{2m}$

d. In the accelerator, work done is fd. Outside, there is no force hence there is no work done. In the region of circular motion with force F, the force is perpendicular to velocity, so no work is done there either. Hence, **total work done is fd.**

END OF PAPER

2014

Part A:

Question 1

For every one green button, there will be two yellow buttons, hence four red buttons, and eight blue buttons. Total is then 15 buttons.

a) $P(b) = 8/15$

b) $P(r) = 4/15$

c) $P(y) = 2/15$

d) $P(g) = 1/15$

Question 2

This is geometric progression with first term 1 and common ratio e^{-x}. Thus, $S_\infty = \frac{u_1}{1-r} = \frac{1}{1-e^{-x}}$.

Sum to infinity would only converge if $|r| < 1$, hence in this case $|e^{-x}| < 1$ which means $e^{-2x} < 1$.

$\rightarrow -2x < \ln(1)\, ; x > 0$

Question 3

a) To solve this, we need to substitute $1 + \sin x = u;\, hence \frac{du}{dx} = \cos x$.

So: $\int_0^{\frac{\pi}{2}} \frac{\cos x}{1+\sin x} dx = \int_0^{\frac{\pi}{2}} \frac{1}{u} du = [\ln u]_0^{\frac{\pi}{2}} = [\ln(1 + \sin x)]_0^{\frac{\pi}{2}} = \ln 2 - \ln 1 = \mathbf{\ln 2}$

b) Using separation of variable, $\int_0^2 \frac{x}{x^2+6x+8} dx = \int_0^2 \frac{2}{x+4} + \frac{-1}{x+2} dx$

$= [2\ln(x + 4) - \ln(x + 2)]_0^2 = \ln 9 - \ln 8 = \mathbf{\ln \frac{9}{8}}$

Question 4

Expanding each term separately, we get:

$(1 + 2x)^4 = 1 + 4(2x) + 6(2x)^2 + 4(2x)^3 + (2x)^4$

$(1 - 2x)^6 = 1 + 6(-2x) + 15(-2x)^2 + 20(-2x)^3 + 15(-2x)^4 + 6(-2x)^5 + (-2x)^6$

The x^7 term will be the product of: $x\ and\ x^6 \rightarrow 8 * (-2)^6 = 512$

$x^2\ and\ x^5 \rightarrow 24 * 6(-2)^5 = -4608$

$x^3\ and\ x^4 \rightarrow 32 * 15(-2)^4 = 7680$

$x^4\ and\ x^3 \rightarrow 16 * 20(-2)^3 = -2560$

So, total coefficient is $512 - 4608 + 7680 - 2560 = \mathbf{1024}$

Question 5

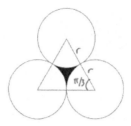

$A = A_{triangle} - 3A_{sector}$;

$A_{triangle} = \frac{1}{2}(2r)^2 \sin\frac{\pi}{3} = \sqrt{3}r^2$

$A_{sector} = \frac{\pi}{3} * \frac{1}{2}r^2 = \frac{\pi r^2}{6}$

$\therefore A = \sqrt{3}r^2 - 3 * \frac{\pi r^2}{6} = \left(\sqrt{3} - \frac{\pi}{2}\right)r^2$

Question 6

Total volume of snowman $= \frac{4}{3}\pi r^3 + \frac{4}{3}\pi (2r)^3 = 12\pi r^3$. This should equal the volume of cylinder. Hence:

$V_{cylinder} = \pi \left(\frac{r}{2}\right)^2 l = 12\pi r^3$;

$\therefore l = 48\,r$

Question 7

The region can be sketched quite straightforwardly. We have a quadratic function, a straight line and limit of y from 0 to 4. The sketch will look as follow:

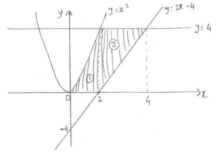

The area can be divided into two:

(1) $\int_0^2 x^2 \, dx = [\frac{1}{3}x^3]_0^2 = \frac{8}{3}$

(2) $\frac{1}{2} *4*2 = 4$

Hence, total area is $\frac{20}{3}$

Question 8

The derivatives can be obtained straightforwardly, and they are shown as follow:

	$x < 0$	$x \geq 0$
$f(x)$	e^x	$e^{-x} + 2x$
$f'(x)$	e^x	$-e^{-x} + 2$
$f''(x)$	e^x	e^{-x}
$f'''(x)$	e^x	$-e^{-x}$

Their plots are as follow:

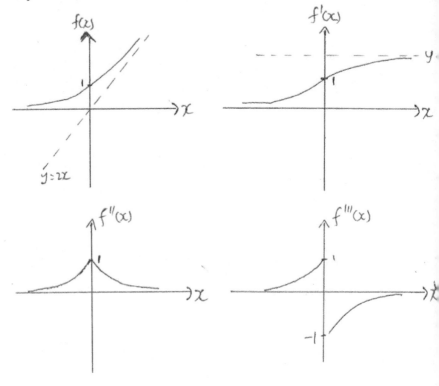

Question 9

The tangents can be illustrated as follow:

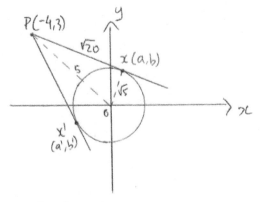

First, we know that $OP^2 = 3^2 + (-4)^2 = 25$; $PX^2 = OP^2 - OX^2 = 25 - 5 = 20$.

Then, we need to find the coordinate of X.

PX: $(a + 4)^2 + (b - 3)^2 = 20$;

$a^2 + 8a + 16 + b^2 - 6b + 9 = 20$;

$a^2 + b^2 + 8a - 6b = -5$ \hfill (1)

OX: $a^2 + b^2 = 5$ \hfill (2)

$(1) - (2)$: $8a - 6b = -10$;

$b = \frac{4a+5}{3}$ \hfill (3)

Substitute b into (1):

$a^2 + (\frac{4a+5}{3})^2 + 8a - 6(\frac{4a+5}{3}) = -5$;

$25a^2 + 40a - 20 = 0$;

$5a^2 + 8a - 4 = 0$;

$(5a - 2)(a + 2) = 0$

$\therefore \boldsymbol{a = \frac{2}{5} or - 2}$

Hence, $\boldsymbol{b = \frac{11}{5} or - 1}$

So, coordinate X is $(\frac{2}{5}, \frac{11}{5})$ and X' is $(-2, -1)$.

The eqn of the lines are then: PX: $y - 3 = \left(\frac{\frac{11}{5}-3}{\frac{2}{5}+4}\right)(x + 4)$;

$\boldsymbol{y = -\frac{2}{11}x + \frac{25}{11}}$

PX': $y - 3 = \left(\frac{-1-3}{-2+4}\right)(x + 4)$;

$\boldsymbol{y = -2x - 5}$

END OF SECTION

Part B:

Question 10: D

Statement i) is wrong as duration of the day depends on the planet's own rotation. Statement ii) is correct because as the planet is further away from the sun, the duration of its year increases (Kepler's 3rd Law). (iii) is not correct as Jupiter is the largest planet. (iv) is incorrect as no. of moons is proportional to the mass of the planet (Uranus and Neptune have less moons than Jupiter and Saturn). Lastly, statement (v) is correct as this change is due to their distance from sun.

Question 11: C

A wave of frequency 100 GHz has a wavelength of $\lambda = \frac{c}{f} = \frac{3 \times 10^8}{10^{11}} = 3\ mm$. This is in a range of **microwave**.

Question 12: A

Using conservation of momentum, since the object's mass is neglected, meaning that the object would still move with the same speed as the ISS. Hence, its centripetal force will still be equal to gravitational force from earth hence it still follows ISS in its orbit.

Question 13

This is pretty straightforward. For parallel arrangement, the total resistance is $R_T = \frac{1}{\frac{1}{R_1} + \frac{1}{R_2} + \cdots + \frac{1}{R_n}}$. Hence:

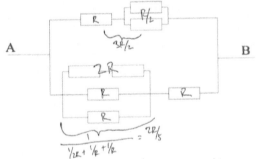

Total resistance is then $R_T = \frac{1}{^2/_{3R} + ^5/_{7R}} = \frac{21R}{29}$

Question 14

For a spring-mass system, we know that oscillation period will be dependent on mass and spring constant in the following relationship $T = 2\pi\sqrt{\frac{m}{k}}$. When two springs are in series, the total spring constant become $k/2$, hence the period will be $T' = 2\pi\sqrt{\frac{m}{k/2}} = \sqrt{2}T$. When the spring are in parallel, the total spring constant become $2k$, hence $T'' = 2\pi\sqrt{\frac{m}{2k}} = \frac{1}{\sqrt{2}}T$.

In other planet, the period of the oscillations will be the same, since period does not depend on g.

Question 15

a). $I = \frac{P}{V}$; $P = mgu = (100)(10)(0.5) = 500\ W$.

$\therefore I = \frac{500}{230} = \frac{50}{23}A$

b). The angular velocity of the reel can be related to the linear velocity of the string   as $v = \omega r$; in this case $v = 3u = 1.5\ m/s$ since it is the $3 -$ pulleys system. Hence: $\boldsymbol{\omega} = \frac{v}{r} = \frac{1.5}{0.025} = \boldsymbol{60\ rads^{-1}}.$

c). Power can also be related to force and velocity with $P = Fv$; $F = \frac{P}{v} = \frac{500}{1.5} =$ **333 N**

Question 16

The threshold voltage of a typical diode is around 0.6 V, below which no current is generated. When positive voltage is applied between C and D, positive current is generated (and vice versa). Hence the sketch will look as follow:

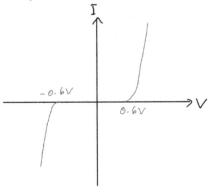

For small signals, hence low voltage, current does not flow through C-D. Then in this case, the amplifier will operate. When there is high voltage, current flows through C-D, not through the amplifier, hence protecting the amplifier.

Question 17

Particle 1 has mass of *2m*, charge *2Q* and speed *v*. Particle 2 has mass of *m*, charge *Q* and speed *v'*. Using conservation of momentum before and after the release:
$0 = 2mv - mv'; \rightarrow v' = 2v$.

Using conservation of energy (electrostatic potential energy = kinetic energy):
$k\frac{(Q)(2Q)}{d} = \frac{1}{2}(2m)(v)^2 + \frac{1}{2}(m)(v')^2 = mv^2 + 2mv^2$;

$\therefore v = \sqrt{\dfrac{2kQ^2}{3md}}$

Question 18

a). Total internal reflection is a phenomenon when waves are entirely reflected, ie. nothing propagated, when striking a medium boundary, going from denser to less dense materials. Critical angle is the angle of incidence above which the total internal reflection occurs. The diagram representation is as follow:

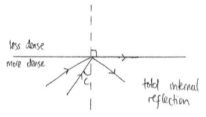

b). starting with $n_1 \sin\theta_1 = n_2 \sin\theta_2$, we know that $\theta_2 = 90^0$ when $\theta_1 = $ critical angle $= c$.

$n_1 \sin(c) = n_2 \sin(90)$;

$\therefore \sin(c) = \dfrac{n_2}{n_1}$

c). In optical fibre, total internal reflection has to happen everywhere along the fibre. When light enters the fibre, we can sketch its propagation as follow:

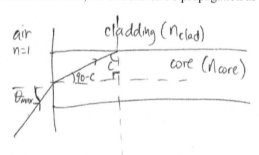

From part b, we know that $\sin(c) = \dfrac{n_2}{n_1} = \dfrac{n_{core}}{n_{clad}}$.

From Snell's law, on the air-fibre boundary, we also know that:

$\sin(\theta_{max}) = n_{core} \sin(90 - c) = n_{core} \cos(c) = n_{core}\sqrt{1 - \sin^2 c} = n_{core}\sqrt{1 - \left(\dfrac{n_{core}}{n_{clad}}\right)^2}$.

$\therefore \boldsymbol{\theta_{max}} = \boldsymbol{\arcsin\left[n_{core}\sqrt{1 - \left(\dfrac{n_{core}}{n_{clad}}\right)^2}\right]}$

d). The water in the tank will refract the light slightly narrower (as water is denser than air), as it enters the tank, before then being widen as it leaves the tank and entering air medium. The illustration can be sketched as follow:

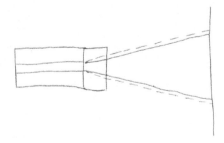

e). As white light contains all visible lights, the image will now be white central spot (light that pass through uninteruptedly), with rainbow around it as each colour light is refracted with different amount.

Question 19

a). We can sketch the system will all the forces displayed as follow:

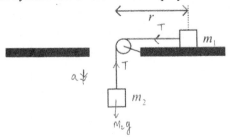

For *m1*: $T = m_1 a;$ (1)

For *m2*: $m_2 g - T = m_2 a;$ (2)

Substituting (1) into (2): $m_2 g - m_1 a = m_2 a;$

$\therefore a = \dfrac{m_2 g}{m_1 + m_2}$

$\therefore T = \dfrac{m_1 m_2 g}{m_1 + m_2}$

b). Assuming the masses are in motion, the equations are modified into the following:

For $m1$: $T - \mu_d m_1 g = m_1 a;$ (1)
For $m2$: $m_2 g - T = m_2 a;$ (2)

Doing the same substitution: $m_2 g - m_1 a - \mu_d m_1 g = m_2 a;$

$$\therefore a = \frac{g(m_2 - \mu_d m_1)}{(m_1 + m_2)}$$

$$\therefore T = m_2 g - m_2 a = m_2 g \left(1 + \frac{\mu_d m_1 - m_2}{(m_1 + m_2)}\right)$$

For motion to exists, T has to be bigger than $\mu_s m_1 g$, and $T < m_2 g$

$\rightarrow \mu_s m_1 g < T < m_2 g$;

$\therefore \mu_s m_1 < m_2$

c). Solving the vertical forces for m_2 (no motion):

$m_2 g - T = 0$; $T = m_2 g$

For $m1$: $T \pm \mu_s m_1 g = m_1 a = m_1 \omega^2 r$;

$r = \frac{m_2 g \pm \mu_s m_1 g}{m_1 \omega^2}$

Depending on r, friction can act radially inwards or outwards, hence the $\pm$ sign. Thus:

$$\therefore r_{min} = \frac{m_2 g - \mu_s m_1 g}{m_1 \omega^2} \; ; r_{max} = \frac{m_2 g + \mu_s m_1 g}{m_1 \omega^2}$$

END OF PAPER

2015

Section A

Question 1

Using Pascal's Triangle, the expansion is $(2x + x^2)^5 = (2x)^5 + (5)(2x)^4(x^2) + (10)(2x)^3(x^2)^2 + 10(2x)^2(x^2)^3 + (5)(2x)(x^2)^4 + (x^2)^5 = \mathbf{32x^5 + 80x^6 + 80x^7 + 40x^8 + 10x^9 + x^{10}}$

Question 2

$\log_2 x + \log_4 16 = 2$;

$\rightarrow \log_2 x + 2 = 2$;

$\rightarrow \log_2 x = 0$;

$\therefore x = 2^0 = 1$

Question 3

This is a geometric series with $a = 1/3$ and common ratio $r = 1/3$. Hence:

$$S_n = \frac{a(1-r^n)}{1-r} = \frac{\frac{1}{3}\left(1-\left(\frac{1}{3}\right)^5\right)}{1-\frac{1}{3}} = \frac{\frac{1}{3}*\frac{242}{243}}{2/3} = \frac{121}{243}$$

$$S_\infty = \frac{a}{1-r} = \frac{\frac{1}{3}}{1-\frac{1}{3}} = \frac{1}{2}$$

Question 4

To solve this integral, we need to do substitution. First, let's say $u = x^2 - 6x + 8$; then $\frac{du}{dx} = 2x - 6$.

Then: $\int_4^6 (2x - 6)[(x - 4)(x - 2)]^{\frac{1}{2}}\, dx = \int_4^6 (2x - 6)(x^2 - 6x + 8)^{\frac{1}{2}}\, dx = \int_4^6 \left(\frac{du}{dx}\right) u^{\frac{1}{2}}\, dx = \int_4^6 u^{\frac{1}{2}}\, du = [\frac{2}{3} u^{\frac{3}{2}}]_4^6$

Substituting u back into the solution:

$\rightarrow [\frac{2}{3}(x^2 - 6x + 8)^{\frac{3}{2}}]_4^6 = \frac{2}{3}\left[(36 - 36 + 8)^{\frac{3}{2}} - (16 - 24 + 8)^{\frac{3}{2}}\right] = \frac{2}{3}\left(8^{\frac{3}{2}}\right) = \frac{32}{3}\sqrt{2}$

Question 5

Putting everything on the left-hand side;

$4x^2 + 8x - 8 - 4mx + 3m = 0;$

$4x^2 + (8 - 4m)x + (3m - 8) = 0$

To have no real solutions, the determinant $D = b^2 - 4ac$ has to be less than zero $(D < 0)$.

→ $D = b^2 - 4ac = (8 - 4m)^2 - 4(4)(3m - 8) < 0$

→ $64 - 64m + 16m^2 - 48m + 128 < 0;$

→ $16m^2 - 112m + 192 < 0;$

→ $m^2 - 7m + 12 < 0;$

→ $(m - 4)(m - 3) < 0;$

Number line:

∴ $3 < m < 4$

Question 6

a). The possible scenarios are TTH, HTT and TTT. Each scenario has a probability of ½ * ½ * ½ = 1/8. Hence, the total probability is **3/8**.

b). The possible scenarios are TTH, THH, HHT and HTT. Total probability is 4/8 = ½.

c). The probability can be expressed as P(all tails) given that one of them is known to be a tail, ie. P(all tails | one tail)

→ P(all tail) = 1/8 ; P(one tail) = 1 – P(all heads) = 1 – 1/8 = 7/8.

Hence, P(all tails | one tail) = (1/8) / (7/8) = 1/7

Question 7

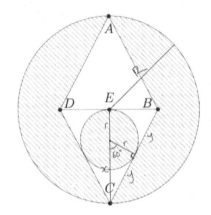

From the sketch above, we know that:

$\cos(60) = \dfrac{r}{x};$

$x = 2r \qquad\qquad (1)$

$\tan(60) = \dfrac{y}{r};$

$y = \sqrt{3}r \qquad\qquad (2)$

$R = x + r = 3r \qquad\qquad (3)$

Area of triangle DCB $= \dfrac{1}{2}(2y)R = \dfrac{1}{2}(2\sqrt{3}r)(3r) = 3\sqrt{3}r^2$

Area of big circle $= \pi R^2 = \pi(3r)^2 = 9\pi r^2$

Shaded area = big circle – 2(triangle) + small circle

$=9\pi r^2 - 2(3\sqrt{3}r^2) + \pi r^2 = \left(5\pi - 3\sqrt{3}\right)2r^2$

Question 8

Finding tangent → differentiate the function: $2(x + 3) + 2(y - 3)\frac{dy}{dx} = 0$;

$\frac{dy}{dx} = \frac{x+3}{3-y}$

At $(1,2)$, $\frac{dy}{dx} = \frac{x+3}{3-y} = 4$

So, **for tangent: *m* = 4** ;

$y = mx + c$;

$2 = 4(1) + c$;

c = -2

For normal:

m = -1/4

$y = mx + c$;

$2 = -\frac{1}{4}(1) + c$;

$c = \frac{9}{4}$

Question 9

Limits:

As $x \to \infty$, $y \to -3$ (from below)

As $x \to -\infty$, $y \to -3$ (from below)

Asymptotes:

$x^2 - 4 = 0$;

$x = \pm 2$

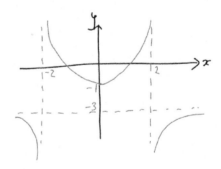

Turning points:

$y = -8(x^2 - 4)^{-1} - 3$;

$\frac{dy}{dx} = 8(x^2 - 4)^{-2}(2x) = 0$

$x = 0$ or $\frac{1}{(x^2-4)^2} = 0$, but the latter has no solutions.

When $x = 0$, $y = -\left(\frac{8}{-4}\right) - 3 = -1$

Hence, the function will look as follow:

$\therefore \{y \geq -1\} \cup \{y < -3\}$

Question 10

We can split these inequalities into two:

$$-1(x-6)^2 < (3x+4)(x-6) \qquad (1)$$

AND

$$(3x+4)(x-6) < (x-6)^2 \qquad (2)$$

(1):
$$-x^2 + 12x - 36 < 3x^2 - 14x - 24$$
$$0 < 4x^2 - 26x + 12$$
$$0 < 2x^2 - 13x + 6$$
$$0 < (2x-1)(x-6)$$

The number line:

$$\rightarrow x < \frac{1}{2} \text{ or } x > 6$$

(2):
$$3x^2 - 14x - 24 < x^2 - 12x + 36$$
$$2x^2 - 2x - 60 < 0$$
$$x^2 - x - 30 < 0$$
$$(x-6)(x+5) < 0$$

The number line:

$$\rightarrow -5 < x < 6$$

Hence, combination of the two limits give:

$$\therefore -5 < x < \frac{1}{2}$$

END OF SECTION

Section B

Question 11

The illustration of the scenario is given below:

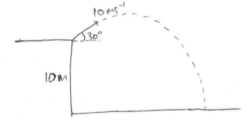

Vertical motion is a constant acceleration (gravity) motion:

$s = 10$ m

u = initial velocity = $-10 \sin(30) = -5$ m/s (-ve as we take downwards as positive)

$v = unknown$

$a = 10$ m/s^2

$t = ?$

$$s = ut + \frac{1}{2}at^2$$
$$10 = -5t + \frac{1}{2}(10)t^2$$
$$0 = 5t^2 - 5t - 10$$
$$0 = t^2 - t - 2$$
$$0 = (t - 2)(t + 1)$$

$t = $ -1 or 2 ; but $t > 0$

$\therefore t = 2s$

Question 12

The scenario can be illustrated as below:

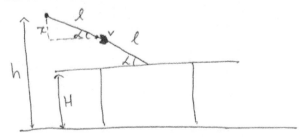

Using similar triangle principle:
$$\frac{x}{l} = \frac{h-H}{2l}$$
$$x = \frac{h-H}{2}$$

Conservation of energy:
$$\frac{1}{2}mv^2 = mgx$$
$$\therefore v = \sqrt{2gx} = \sqrt{g(h-H)}$$

Question 13

Let's first sketch the sun-moon-earth system:

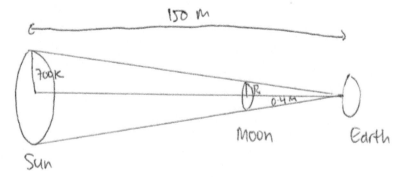

We can then use the similar triangles principle:
$$\frac{R}{700} = \frac{0.4}{150}$$
$$R = \frac{700*0.4}{150} = \frac{28}{15}$$
$$\therefore Radius \approx 2000 \; km$$

Question 14
The sketch is straightforward as below:

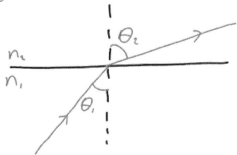

$n_1 sin\theta_1 = n_2 sin\theta_2$
Total internal reflection occurs when $n_2 < n_1$
For $\theta_1 > \theta_c$, where:
$n_1 sin\theta_c = n_2$
$\therefore \boldsymbol{\theta_c = sin^{-1}(\frac{n_2}{n_1})}$

Question 15

As the mass does not move, both horizontal and vertical forces have to balance out.

Horizontal:
$A = C * \cos(45)$
$C = \frac{2A}{\sqrt{2}} = \sqrt{2}A$

Vertical:
$B = C * \cos(45)$
$\therefore \boldsymbol{B = (\sqrt{2}A)\left(\frac{\sqrt{2}}{2}\right) = A}$

Question 16
Let's call the speed after collision of 2kg mass is v_1 and of 1 kg mass v_2.

2kg →1 m s⁻¹   1 kg ⇒   2kg →v_1 1 kg →v_2

Conservation of momentum:
$$2(1) + 0 = 2v_1 + v_2$$
$$2 = 2v_1 + v_2 \qquad (1)$$
For elastic collision, coefficient of restitution $e = 1$. Hence:
$$v_2 - v_1 = 1(1 - 0)$$
$$v_2 - v_1 = 1 \qquad (2)$$

(1)-(2):
$$1 = 3v_1$$
$$\therefore v_1 = 0.3 \ ms^{-1}$$
(2):
$$v_2 = 1 + v_1$$
$$\therefore v_2 = 1.3 \ ms^{-1}$$

Question 17

First, find the frequency:
$$f = \frac{v}{\lambda} = \frac{2}{10} = 0.2 \ Hz$$
$$\omega = 2\pi f = 0.4\pi$$
Hence, max. vertical velocity of the boat $= \omega A = (0.4\pi)(0.5) = \mathbf{0.2\pi \ ms^{-1}}$

Question 18
a)   Speed = volume flow rate/cross-section area = x/A, in m/s (if A is in m²)

b) $F = \frac{\Delta p}{\Delta t} = \frac{m\Delta v}{\Delta t}; \ \frac{m}{\Delta t} = \frac{\rho V}{\Delta t} = \rho x \rightarrow F = \rho x \Delta v$

i) If the water falls to the ground when it hits the wall, meaning its final speed after collision is $0 \rightarrow \Delta v = \frac{x}{A}$

$$\therefore F = \rho x * \frac{x}{A} = \frac{\rho x^2}{A}$$

ii) If it rebounds $\rightarrow \Delta v = \frac{x}{A} - \left(-\frac{x}{A}\right) = \frac{2x}{A}$

$$\therefore F = \rho x * \frac{2x}{A} = \frac{2\rho x^2}{A}$$

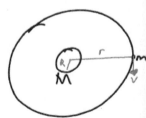

Question 19

For stable orbit; gravitational potential force = centripetal force

$$\frac{GMm}{r^2} = \frac{mv^2}{r}$$

$$v = \sqrt{\frac{GM}{r}}$$

$$g = \frac{GM}{r^2} = 10 \; ;$$

For orbit around equator at sea level $\rightarrow r = R = 6.4 * 10^6$

Hence: $v = \sqrt{\frac{GM}{r}} == \sqrt{\frac{GMr}{r^2}}$, in the case where $r = R \rightarrow v = \sqrt{\frac{GMR}{R^2}} == \sqrt{gR} =$
$\sqrt{(10)(6.4 * 10^6)} = 8000 \; ms^{-1}$

In reality, it would come across mountains hence will make it difficult for the satellite to maintain a circular orbit.

Question 20

Start with Planck – Einstein relation: $E = hv = \frac{hc}{\lambda}$

a). shortest wavelength corresponds to highest energy emitted $\rightarrow$ De-excitation happen from $n = 10$ to $n = 1$.

$$\frac{hc}{\lambda} = E = -\frac{R}{(10)^2} - \left(-\frac{R}{(1)^2}\right) = \frac{99R}{100} \rightarrow \lambda = \frac{100hc}{99R}$$

b). longest wavelength corresponds to lowest energy emitted $\rightarrow$ De-excitation happen from $n = 10$ to $n = 9$.

$$\frac{hc}{\lambda} = E = -\frac{R}{(10)^2} - \left(-\frac{R}{(9)^2}\right) = \frac{19R}{8100} \rightarrow \lambda = \frac{8100hc}{19R}$$

c). Emission lines correspond to possible de-excitation routes, which can be illustrated as below:

If electron is in $n = 10$, then there are 9 possible de-excitation routes.
If electron is in $n = 9$, then there are 8 possible de-excitation routes. And so on…

Hence $\rightarrow 9 + 8 + 7 + \ldots + 1 = \mathbf{45}$

Question 21

The resistors with their values can be seen below:

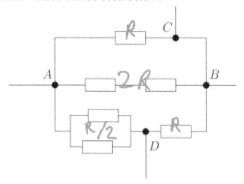

a). Total resistance between A and B:

$$\frac{1}{R_{A-B}} = \frac{1}{R} + \frac{1}{2R} + \frac{1}{3R/_2} = \frac{13}{6R}$$

$$\therefore R_{A-B} = \frac{6R}{13}$$

b). Potential across BD will be a fraction of AB (V), proportional to R relative to $3R/2$ (potential divider model):

$$V_{BD} = \frac{R}{3R/_2} * V = \frac{2V}{3} \; ; P = \frac{V^2}{R} = \frac{\left(2V/_3\right)^2}{R} = \frac{4V^2}{9R}$$

c). To calculate total resistance between C and D, it is easier to redraw the circuit so that C and D are the end-terminals:

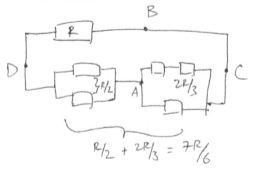

From this figure, it is then straightforward that:

$$\frac{1}{R_{CD}} = \frac{1}{R} + \frac{6}{7R} = \frac{13}{7R} \; ; R_{CD} = \frac{7R}{13}$$

END OF PAPER

2016

Section A

Question 1

To solve this, we need to use product and chain rules:

If $f(x) = g(x)h(x)$;

$f'(x) = g'(x)h(x) + g(x)h'(x)$

Applying this to the question:

$f'(x) = \sin(x^2) + x\cos(x^2)(2x)$

$= \mathbf{\sin(x^2) + 2x^2\cos(x^2)}$

Question 2

$\sqrt{3}\tan^2\theta - 2\tan\theta - \sqrt{3} = 0$;

$(\sqrt{3}\tan\theta + 1)(\tan\theta - \sqrt{3}) = 0$

$\tan\theta = -\frac{1}{\sqrt{3}}$ or $\tan\theta = \sqrt{3}$

$\therefore \boldsymbol{\theta = \frac{5\pi}{6}, \frac{11\pi}{6}, \frac{\pi}{3}, \frac{4\pi}{3}}$

Question 3

From the 1st equation:

$\log_4\left(\frac{64^x}{16^y}\right) = 13$;

$\log_4 64^x - \log_4 16^y = 13$;

$\log_4 4^{3x} - \log_4 4^{2y} = 13$;

$3x\log_4 4 - 2y\log_4 4 = 13$;

$\therefore 3x - 2y = 13$

2nd equation:

$\log_{10} 10^x + \log_3 3^y = 1$;

$x\log_{10} 10 + y\log_3 3 = 1$;

$\therefore x + y = 1$

Substitute y into x:

$3x - 2(1 - x) = 13$;

$\therefore \boldsymbol{x = 3}$

And $\boldsymbol{y = -2}$

Question 4

The expansion will be:

$$(^{12}_0 C)(x)^{12}\left(-\frac{1}{x^2}\right)^0$$

$$+ (^{12}_1 C)(x)^{11}\left(-\frac{1}{x^2}\right)^1$$

$$+ \ldots$$

$$+ (^{12}_4 C)(x)^8\left(-\frac{1}{x^2}\right)^4 \rightarrow \text{independent of } x \text{ , with coefficient } ^{12}_4 C$$

$$\therefore {}^{12}_4 C = \frac{12!}{4!8!} = \frac{12 \times 11 \times 10 \times 9}{4 \times 3 \times 2 \times 1} = \textbf{495}$$

Question 5

Let's start with 4-digit number. We can start the 1st digit either with 5, 6 or 7 (ie. 5xxx, 6xxx, 7xxx).

If the first digit is 5, the 2nd digit can take any of the remaining numbers (4 options, ie. 3, 4, 6 and 7). The third digit can then be any of the remaining number again (now 3 options left, ie. if the 2nd digit is 3, then the remaining options are 4, 6 and 7). The fourth digit has two possible options. Hence the possible scenario, for 1st digit of 5 is 4 x 3 x 2 = 24.

Total possible scenarios for 4-digit number is then 3 x 24 = 72.

For 5-digit number, the 1st digit can be any of the options (5 options), the 2nd digit can be whatever left (4 options), and so on. Hence the total scenario for 5-digit number is 5 x 4 x 3 x 2 x 1 = 120.

$\therefore$ total possible scnearios $= 72 + 120 = \textbf{192}$

Question 6

The scenario can be expressed as follow:

Month:	0	1	2	3	$\ldots$
Twig:	0	1	2	4	$\ldots$
Leaves:	0	2	4	8	$\ldots$

We can see that the no. of leaves follow geometric progression with first term $a = 2$ and common ratio $r = 2$.

$$S_{10} = \frac{a(r^{10}-1)}{r-1} = \frac{2(2^{10}-1)}{2-1} = \textbf{2046}$$

Question 7

The sequence has to be 6,5,4,3,3,3,3,2,2,1 (10 throws).

a). If it is only thrown for 8 times, then obviously it is not possible (zero probability).

b). For 10 throws, then he must get everything right → probability is $(1/6)^{10}$

c). For 12 throws, the scenario can then be one of the followings:

-1st and 2nd throw can be anything (P = 1), 3rd till 12th throw has to follow the sequence $(P = (1/6)^{10})$ → $P = (1/6)^{10}$

-1st throw can be anything, 2nd till 11th throw follow the sequence, 12th throw can be anything → $P = (1/6)^{10}$

-1st till 10th throw follow the sequence, 11th and 12th throws can be anything → $P = (1/6)^{10}$

∴ total probability $= 3 * \left(\frac{1}{6}\right)^{10} = \frac{3}{6^{10}}$

Question 8

Octagon with parallel length x has the area of square with sides x minus 4 little right-angle triangles of the edges.

We know that s and d relate with x in the following way:

$x = s + 2d$, but by Pythagoras we also know that $s^2 = 2d^2 ; d = \frac{s}{\sqrt{2}}$

Hence: $x = s + \sqrt{2}s = \left(1 + \sqrt{2}\right)s$ (1)

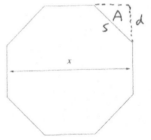

Area of octagon $= x^2 - 4A$

$= x^2 - 4(\frac{1}{2}d^2)$

$= x^2 - 4\left[\frac{1}{2}\left(\frac{s}{\sqrt{2}}\right)^2\right]$

$= x^2 - 4\left[\frac{1}{4}s^2\right]$

$= x^2 - s^2$; substituting (1):

$= x^2 - \left(\frac{1}{1+\sqrt{2}}\right)^2 x^2$

$= \frac{2x^2}{1+\sqrt{2}}$

For area $= \frac{2x^2}{1+\sqrt{2}} = \pi r^2$;

∴ $x = \sqrt{\frac{\pi r^2(1+\sqrt{2})}{2}}$

Question 9

Multiplying both sides by x^2 (to make sure it is always positive):

$5x^2 - 3x^3 < 2x$

$0 < x(3x^2 - 5x + 2)$

$0 < x(3x - 2)(x - 1)$

The number line:

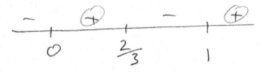

$\therefore \left\{0 < x < \frac{2}{3}\right\} \cup \{x > 1\}$

Question 10

Sketch of the curves:

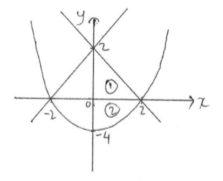

Area (1):

$\frac{1}{2}(2)(2) = 2$

Area (2):

$\int_0^2 (x^2 - 4)dx = [\frac{x^3}{3} - 4x]_0^2 = \frac{8}{3} - 8 = -\frac{16}{3}$ (-ve value as it is below the x-axis).

Since the area is symmetric with respect to the y-axis, the total area is then $= 2 \times (2 + 16/3) = \mathbf{44/3}$

Question 11

We need to find dr/dt. We can use chain rule to do it.

$$h(t) = h_0 - \alpha t \; ;$$

$$\frac{dh}{dt} = -\alpha$$

Volume is constant:

$$V = \pi r^2 h = constant$$

$$h = \frac{V}{\pi r^2} = \frac{V}{\pi} r^{-2}$$

$$\frac{dh}{dr} = (-2)\frac{V}{\pi} r^{-3} = \frac{-2V}{\pi r^3}$$

$$\frac{dr}{dt} = \frac{dr}{dh} * \frac{dh}{dt} = \frac{\pi r^3}{-2V} * (-\alpha) = \frac{\pi r^3 \alpha}{2V}$$

We need to substitute r in terms of V and h_0:

$$r = \sqrt{\frac{V}{\pi h}} = \sqrt{\frac{V}{\pi(h_0 - \alpha t)}}$$

$$\rightarrow \frac{dr}{dt} = \frac{\pi \left(\frac{V}{\pi(h_0 - \alpha t)}\right)^{\frac{3}{2}} \alpha}{2V} = \frac{\alpha}{2}\sqrt{\frac{V}{\pi}\frac{1}{(h_0 - \alpha t)^3}}$$

From the above equation, we can tell that the rate increases with time.

END OF SECTION

Section B

Question 12

We first need to realise that the height of the ball is proportional to its energy, ie. $E_p = mgh = E_k$.

After the first bounce, the kinetic $\equiv$ potential energy $\equiv$ height become $\frac{3}{4}$ of initial.

After the second bounce, the height become $\frac{3}{4} \times \frac{3}{4} = (\frac{3}{4})^2$ of its initial.

After the n bounce, the height become $(\frac{3}{4})^n$ of its initial.

Hence, $(\frac{3}{4})^n < \frac{1}{4}$; $n = 5$ is the smallest integer that satisfies this inequality. **So it must bounce 5 times.**

Question 13

a). For an ideal wire, resistance is proportional with temperature.

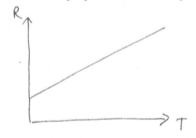

b). Filament light bulb's resistance increases with temperature, though not linearly. Instead, its variation is as follow:

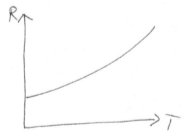

c). A thermistor's resistance decreases with temperature as follow:

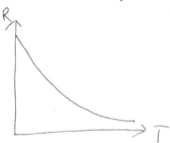

Question 14

Kepler's 3rd Law $\rightarrow T^2 \propto r^3$

Given that $\frac{r_E}{r_I} = 1.6$; $\left(\frac{r_E}{r_I}\right)^3 = (1.6)^3 = \left(\frac{16}{10}\right)^3$

Hence, $\frac{T_E}{T_I} = \left(\frac{16}{10}\right)^{\frac{3}{2}} = \left(\frac{4}{\sqrt{10}}\right)^3 = \frac{64}{10\sqrt{10}} = \frac{6.4}{\sqrt{10}}$

What is $\sqrt{10}$ in decimals? We can estimate this. Try $3.2^2 = 10.24$, $3.1^2 = 9.61$. $\rightarrow$ $3.2^2 = 3.2$ (1 dp).

$\therefore \frac{T_E}{T_I} = \frac{6.4}{3.2} = \mathbf{2.0}$

Question 15

We can use vector to solve this. The scenario can be illustrated as follow:

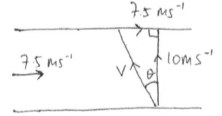

The above vector triangle is proportional to the following:

From the above simplified triangle, we can work out $v \rightarrow v = 5*(7.5/3) = \mathbf{12.5}$ **m/s.**

$\tan \theta = \frac{7.5}{10} = \frac{3}{4};$

$\theta = \arctan\left(\frac{3}{4}\right)$

Relative to the flowing water, the angle is then $\frac{\pi}{2} + \mathbf{arctan}\left(\frac{3}{4}\right)$

Question 16

If q was x away from Q_1, then it is $(a-x)$ away from Q_2. Then:

$$F_1 = kqQ_1/x^2; \quad F_2 = kqQ_2/(a-x)^2$$

Then, $F_1 = F_2$

$$\frac{kqQ_1}{x^2} = \frac{kqQ_2}{(a-x)^2}$$

Take $Q_2 = nQ_1$:

$$\frac{Q_1}{x^2} = \frac{nQ_1}{(a-x)^2}$$

$$(a^2 - 2ax + x^2) = nx^2$$

$$(1-n)x^2 - 2ax + a^2 = 0;$$

Using a,b,c formula:

$$x = \frac{2a \pm \sqrt{4a^2 - 4a^2(1-n)}}{2(1-n)} = \frac{2a \pm 2a\sqrt{n}}{2(1-n)} = \frac{a \pm a\sqrt{n}}{(1-n)} = \frac{a(1 \pm \sqrt{n})}{(1+\sqrt{n})(1-\sqrt{n})}$$

$$x = \frac{a}{1+\sqrt{n}} \text{ or } \frac{a}{1-\sqrt{n}}$$

For like charges, x has to be positive. Hence: $x = \dfrac{a}{1+\sqrt{n}}$

When $Q_1 = Q_2 \rightarrow n = 1 \rightarrow x = a/2$

When $Q_1 \neq Q_2; n = \dfrac{Q_2}{Q_1}$

$$x = \frac{a}{1 + \sqrt{\frac{Q_2}{Q_1}}}$$

Question 17

a). Work is the area under curve for force-displacement diagram. Hence, work from $x = 5$ cm to 0 cm is $= \frac{1}{2}$ **(0.05)(10) = 0.25 J**

b). The work done = kinetic energy $= \frac{1}{2} mv^2$. Hence $\rightarrow v = [(0.25)*2/(0.02)]^{1/2} =$ **5 m/s**

c). From the diagram, we know that the mathematical relationship between force and displacement is $F = -kx$. This is simple harmonic motion, with acceleration $a = -\omega^2 x$.

$$\rightarrow \omega^2 = -\frac{a}{x} = \frac{-F/m}{x} = \frac{-\frac{(-12)}{0.02}}{0.06} = 10000 \; ; \; \omega = 100$$

$$T \text{ (period)} = \frac{2\pi}{\omega} = \frac{\pi}{50} = \mathbf{0.06 \; s}$$

Hence, the mass moves with simple harmonic motion, with period of 0.06 s and amplitude of 0.05 m.

Question 18

Unit of force = N = kg m s^{-2}

Unit of radius = m

Unit of velocity = m s^{-1}

Unit of coeff. Of viscosity = kg m^{-1} s^{-1}

$$kgms^{-2} = (m)^x(kgm^{-1}s^{-1})^y(ms^{-1})^z = (m)^{x-y+z}(kg)^y(s)^{-y-z}$$

Matching the power of each dimension on both sides:

Mass (kg):   **1 = y**

Length (m): $1 = x - y + z$ (2)

Time (s): $-2 = -y - z$ (3)

Substitute y into (3): $-2 = -1 - z$; $z = 1$

Substitute y and z into (2): $1 = x - 1 + 1$; $x = 1$

Question 19

a). Energy of the light will be used to excite the electron out of the metal surface, then surplus energy is used to accelerate it (ie. kinetic energy). Hence:

$$\frac{hc}{\lambda} = \phi + \frac{1}{2}mv_{mas}^2$$

$$\frac{(6*10^{-34})(3*10^8)}{625*10^{-9}} = 1.6 * 10^{-19} + \frac{1}{2}(1*10^{-30})v^2 \ ;$$

$$5.7 * 10^{-19} - 3.2 * 10^{-19} = 10^{-30}v^2 \ ;$$

$$v^2 = \frac{(2.5*10^{-19})}{10^{-30}} = 25 * 10^{10} \ ;$$

$$\therefore v = 5 * 10^5 \ ms^{-1}$$

b). Energy provided by the potential to an electron $= 5 * 10^3 \ x \ 1.6 * 10^{-19} = 8 * 10^{-16} J$

We can then ignore the initial v since it will be negligibly small compared to the final speed:

$$8 * 10^{-16} = \frac{1}{2} * (10^{-30})v^2;$$

$$v^2 = 16 * 10^{14};$$

$$\therefore v = 4 * 10^7 \ ms^{-1}$$

Question 20

We should not be confused with what seem to be a complex circuit – first, we know that the two resistor at top, BC, and two resistors at the bottom are in parallel with voltage = 84V across them.

The so-called "junctions" (point between the top resistors, point between B and C) have the same potential since the two resistors have the same resistance, and so do the heaters. **Hence, heater A won't heat since there is no pd. across it.**

Heater B and C will each has 42 V across it – power is then $P = \dfrac{V^2}{R} = \dfrac{(42)^2}{6} = 294\ W$.

The amount of energy needed to heat 1 kg of water from 20 to 27C is $mc\Delta T = (1)(4200)(27 - 20) = 29400\ J$.

The time taken is then $t = \dfrac{E}{P} = \dfrac{29400}{294} = \mathbf{100\ s}$

Question 21

The diagram of the incident (and refracted) light can be drawn as follow:

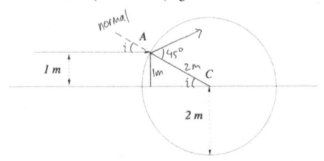

a). We can see that $\sin i = \dfrac{1}{2}$; and $n = \dfrac{\sin r}{\sin i} = \dfrac{\sin 45}{1/2} = \sqrt{2} = \mathbf{1.4}$

b). We want to find the critical angle for total internal reflection. Refractive index $n = \sqrt{2}$

$\rightarrow \sqrt{2} = \dfrac{\sin 90}{\sin \theta_c}$;

$\sin \theta_c = \dfrac{1}{\sqrt{2}}$

$\therefore \theta_c = \mathbf{45^o}$

Hence, beam should be directed at an angle greater than 45°

END OF PAPER

2017

Question 1: D

Using chain rule: $\frac{d(uv)}{dx} = v\frac{du}{dx} + u\frac{dv}{dx}$.

In this case, $u = 2x$; $v = \cos x$.

Hence, $\frac{du}{dx} = 2$; $\frac{dv}{dx} = -\sin x$

$\rightarrow \frac{d(2x\cos x)}{dx} = \cos x \cdot 2 + 2x \cdot (-\sin x) = \mathbf{2\cos x - 2x\sin x}$

Question 2: A

Divide both sides of the equation by 2 $\rightarrow x^2 - x - 6 = 0$

Factorise it: $(x + 2)(x - 3) = 0$

Question 3: E

For $S = \sum_0^n x^{-i}$;

$xS = \sum_1^{n+1} x^i = S - 1 + x^{n+1}$

$\rightarrow S = \frac{x^{n+1}-1}{x-1}$

In this case, $x = -e^{-1}$

So: $\sum_0^{10} x^n = \frac{-e^{-11}-1}{-e^{-1}-1} = \frac{1+e^{-11}}{1+e^{-1}}$

Question 4: D

Taking log of both sides:

$\log(a^{3-x}b^{5x}) = \log(a^{x+5}b^{3x})$

$\rightarrow (3-x)\log a + 5x\log b = (x+5)\log a + 3x\log b$

$\rightarrow 2x\log b - 2x\log a = 2\log a$

$\rightarrow x = \frac{\log a}{\log b - \log a}$

Question 5: B
Let's sketch the functions so that we can check whether the area will cancel out or not:

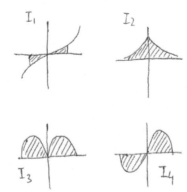

From the sketches above, we can see that I_1 **and** I_4 will have zero total area since the negative area equals the positive area.

Question 6: C
We can check the asymptotes of all the possible functions:
Option A has asymptotes at $x = 1$ and -3. Limit to $x = -3$ from the left will be negative; however, the graph is positive.
Option B has asymptotes at $x = 3$, -1, hence cannot represent the graph.
Option C has asymptotes at $x = 1$ and -3. When $x = 0$, A = -1/3 so this can represent the graph.
Option D has asymptotes at $x = 1, -1, -3$ (extra asymptotes from what is required), so not representing the graph.
Option E has asymptotes at $x = 3$, -3, hence not the graph either.

Question 7: C
The same thing would happen on the moon as would happen on earth, or any other body in that matters. Moon has its own gravitational pull so the ball would fall to the surface of the moon eventually.

Question 8: E
You just need to remember this. From shortest to longest wavelength (highest to lowest energy):
X-ray, ultraviolet, visible, infrared, radio

Question 9: B

The overall resistance of the parallel arrangement:

$$\frac{1}{R_p} = \frac{1}{R} + \frac{1}{2R} \; ; R_p = \frac{2R}{3}$$

$$R_{total} = \frac{2R}{3} + R = \frac{5R}{3}$$

$$\therefore I = \frac{V}{R_{total}} = \frac{V}{\frac{5R}{3}} = \frac{3V}{5R}$$

Question 10: A

Capacitance is directly proportional to surface area $(c = \frac{A\varepsilon}{d})$. So, if the area is halved, capacitance is halved.

Question 11: B

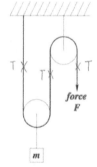

Evaluate newton's law on the both pulleys:

$F = T$ (1)

$mg = 2T$ (2)

Hence, **$F = mg/2$**

Question 12: A

By conservation of energy: $qV = \frac{1}{s}mv^2; \rightarrow mv = \frac{2qV}{v}$

Force is constant, hence acceleration is constant → straight line with negative slope on the speed-time graph, with y-intercept of the initial speed. Distance $d =$ area of speed-time graph = ½ vt → $\frac{2}{v} = \frac{t}{d}$

Substitute $2/v$ into the first equation:

$mv = \frac{qVt}{d}$ and this is the initial momentum.

Question 13
Using pascal's triangle:
$$(3 + 2x)^5 = 3^5 + 5(3)^4(2x) + 10(3)^3(2x)^2 + 10(3)^2(2x)^3 + 5(3)(2x)^4 + (2x)^5$$
$$= 243 + 810x + 1080x^2 + 720x^3 + 240x^4 + 32x^5$$

Question 14
To be busy, $P(A) = 0.5$, $P(B) = 0.75$ and $P(C)= 0.2$. So, **$P(A \cap B \cap C) = 0.5*0.75*0.2 = 0.075$**
$P(\text{all free}) = (1-P(A)) * (1-P(B)) * (1-P(C)) = 0.5*0.25*0.8 = 0.1$

Question 15
To cause mass m to move, the elastic force from the spring must exceed the mass m's static friction; $> \mu_s mg$, but $T = kx$.
Hence: $x > \mu_s mg$; $x > \frac{\mu_s mg}{k}$

Question 16

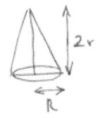 

Volume of cone $= \frac{2}{3}\pi R^2 r$
Volume of sphere $= \frac{4}{3}\pi r^3$

If the volumes are equal: $\frac{2}{3}\pi R^2 r = \frac{4}{3}\pi r^3$;
→ $R^2 = 2r^2$
∴ $R = r\sqrt{2}$

Question 17

$m\frac{dv}{dt}$ has the unit of kg m s⁻². Hence, α has to take the unit of **kg.m⁻¹** since v^2 has the unit of m²s⁻².

Terminal velocity happens when $\frac{dv}{dt} = 0$. → $mg - \alpha v^2 = 0$; $v = \sqrt{\frac{mg}{\alpha}}$

Potential energy lost $= mgh$

Kinetic Energy gained $= \frac{1}{2}mv_{term}^2 = \frac{1}{2}m\left(\frac{mg}{\alpha}\right)$

∴ **work done by drag** $= mgh - \frac{1}{2}\frac{m^2g}{\alpha} = mg\left(h - \frac{m}{2\alpha}\right)$

Question 18

Speed = wavelength x frequency $= \frac{\lambda_1\omega_1}{2\pi} = \frac{\lambda_2\omega_2}{2\pi}$

$y_1 + y_2 = 2A\left(\cos\left(\pi x\left(\frac{1}{\lambda_1} + \frac{1}{\lambda_2}\right) - \frac{\omega_1+\omega_2}{2}t\right)\right)\left(\cos\left(\pi x\left(\frac{1}{\lambda_1} - \frac{1}{\lambda_2}\right) + \frac{\omega_2-\omega_1}{2}t\right)\right)$

$\qquad = 2AC_1C_2$ where wavelengths of $C_1 = L_1$, $C_2 = L_2$

$\pi x\left(\frac{\lambda_1+\lambda_2}{\lambda_1\lambda_2}\right) = \frac{2\pi x}{L_1}$; → $L_1 = \frac{2\lambda_1\lambda_2}{\lambda_1+\lambda_2}$

$\pi x\left(\frac{\lambda_2-\lambda_1}{\lambda_1\lambda_2}\right) = \frac{2\pi x}{L_2}$; → $L_2 = \frac{2\lambda_1\lambda_2}{\lambda_2-\lambda_1}$

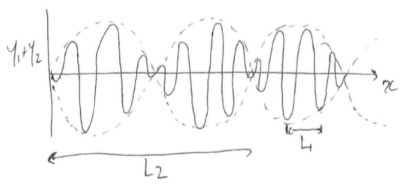

Frequency you hear = speed/$L_1 = \frac{\lambda_1\omega_1}{2\pi}\left(\frac{\lambda_1+\lambda_2}{\lambda_1\lambda_2}\right) = \frac{\omega_1}{2\pi}\left(\frac{\lambda_1+\lambda_2}{\lambda_2}\right)$

Distance $= \frac{L_2}{2} = \left(\frac{\lambda_1\lambda_2}{\lambda_2-\lambda_1}\right)$

Question 19

$y = 0$ when $\cos \omega t = \frac{\sqrt{3}}{2}$, $\sin \omega t = \pm \frac{1}{2}$

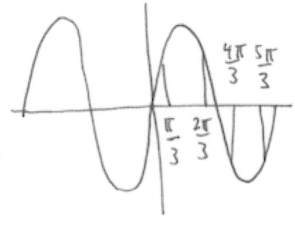

$$\omega t = (2n + \frac{i}{3})\pi$$

Where n is a (+ or -) interger and $i \in \{1, 2, 4, 5\}$

$$\therefore x = \alpha\left(\left(2n + \frac{1}{3}\right)\pi - \frac{1}{2}\right) \text{ or } \alpha\left(\left(2n + \frac{2}{3}\right)\pi - \frac{1}{2}\right) \text{ or } \alpha\left(\left(2n + \frac{4}{3}\right)\pi + \frac{1}{2}\right)$$

$$\text{or } \alpha\left(\left(2n + \frac{5}{3}\right)\pi + \frac{1}{2}\right) \text{ where } n \text{ is an interger}$$

Question 20

In the 2 bodies system:

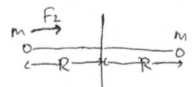

$$F_2 = \frac{Gm^2}{(2R)^2} = \frac{Gm^2}{4R^2}$$

3 bodies system:

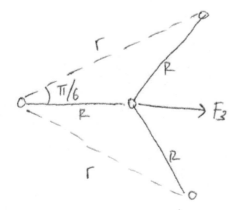

$$r = 2R\cos\frac{\pi}{6} = R\sqrt{3}$$

$$F_3 = 2\left(\frac{Gm^2}{r^2}\cos\frac{\pi}{6}\right) = 2\frac{Gm^2}{\left(R\sqrt{3}\right)^2}\frac{\sqrt{3}}{2} = \frac{Gm^2}{R^2\sqrt{3}}$$

$$\frac{mv_2^2}{R} = F_2 , \frac{mv_3^2}{R} = F_3 \rightarrow \frac{v_3^2}{v_2^2} = \frac{F_3}{F_2} = \frac{4}{\sqrt{3}}$$

$$\therefore v_3 = v_2 \left(\frac{4}{\sqrt{3}}\right)^{\frac{1}{2}}$$

Question 21

To solve this expression, we can use Leibniz's rule:

$$\frac{d}{dt}\int_0^{2t^2}(xt)^4\,dx = \left((2t^2)t\right)^4\frac{d}{dt}(2t^2) - \left((0)t\right)^4\frac{d}{dt}(0) + \int_0^{2t^2}\frac{\partial}{\partial t}(xt)^4\,dx$$

$$= 16t^8.t^4.4t - 0 + \int_0^{2t^2}4t^3x^4\,dx = 64t^{13} + [\frac{4t^3.x^5}{5}]_0^{2t^2} = 64t^{13} + \frac{4t^3.32t^{10}}{5} =$$

$$\frac{448t^{13}}{5}$$

Question 22

We can rewrite the equations of the circles into the following (remind yourself on how to do this):

C_1: $(x + 3)^2 + (y - 2)^2 = 1^2$
C_2: $(x - 5)^2 + (y - 1)^2 = 5^2$

The sketch of the circles and their common tangents is shown below:

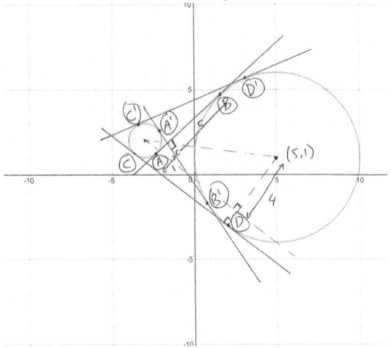

There are 4 tangents between these two cirlces: AB, A'B', CD and C'D'.

From the sketch, it is clear to see that **AB = A'B' = sum of the radius of the circles = 5+1 = 6**

To find CD, we need to use Pythagoras:
$CD^2 + 4^2 = (5 - (-3))^2 + (1 - 2)^2$
$CD^2 + 16 = 64 + 1$
$CD^2 = 49$
$\therefore CD = C'D' = 7$

Question 23

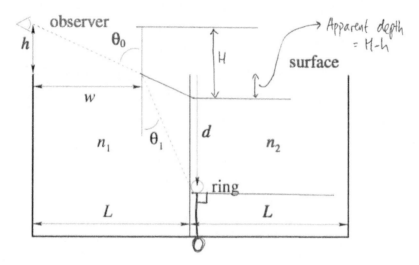

Let θ_1 be the minimum angle, below which the ray from the ring moves vertically along the n_1-n_2 boundary.

Air to n_1:
$$\sin \theta_0 = n_1 \sin \theta_1 \tag{1}$$

Secondly;
$$n_2 \sin \frac{\pi}{2} = n_1 \sin \left(\frac{\pi}{2} - \theta_1 \right) \rightarrow n_2 = n_1 \cos \theta_1 \tag{2}$$

Square the (1) equation:
$$\sin^2 \theta_0 = n_1^2 \sin^2 \theta_1 \ ; \ \text{but} \ \sin^2 \theta_1 = 1 - \cos^2 \theta_1 = 1 - \left(\frac{n_2}{n_1} \right)^2$$
$$\rightarrow \sin^2 \theta_0 = n_1^2 \left(1 - \left(\frac{n_2}{n_1} \right)^2 \right) = n_1^2 - n_2^2 = \mu \ (\text{for brevity})$$

Furthermore, from the figure geometry we know that:
$$\sin^2 \theta_0 = \frac{L^2}{H^2 + L^2} \rightarrow H^2 = L^2 \frac{1 - \mu}{\mu}$$
$$\therefore \ Max. \, apparent \, depth = H - h = \sqrt{\frac{L^2(1-\mu)}{\mu}} - h \ ; \ \text{where} \ \mu = n_1^2 - n_2^2$$

END OF PAPER

2018

Question 1: C

The differences of the sequence are: 1,4,5,9... This is similar to the Fibonacci sequence (each subsequent term is the sum of the two previous ones). The next difference would therefore be 5+9=14, meaning the next term would be 23+14=37.

Question 2: E

All could be trajectories except 3 which is a circular orbit but the central star doesn't align with the centre of the orbit (and therefore gravity wouldn't be acting in the correct direction for part of the orbit).

Question 3: E

Convert into common units.

A: $CVm^{-1} = Jm^{-1} = Nmm^{-1}$. This is a measure of force.

B: $ATm = A(Wbm^{-2})m = A(kgs^{-2}A^{-1})m = kgms^{-2}$. This is force.

C: $kgms^{-2}$ This is a unit of force (eg. $F = ma$)

D: Jm^{-1} This is a unit of force (see A).

E: $Cms^{-1} = Asms^{-1} = Am$ This is the odd one out

Question 4: C

Draw a diagram.

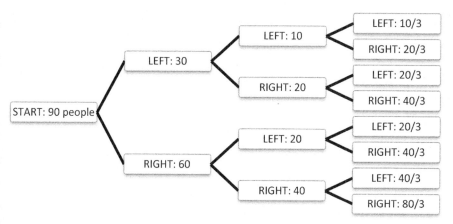

Adding up the different options gives:

LLL = 10/3

2xL, 1xR = 20/3+20/3+20/3=20

2xR, 1xL = 40/3+40/3+40/3=40 (This is the most likely)

RRR = 80/3

Question 5: C

Consider this as a geometric progression, with the volume of the first cup defined as 'V': $V + \alpha V + \alpha^2 V + \cdots \alpha^{n-1} V + \alpha^n V \leq 3V$

Factorising out V gives $1 + \alpha + \alpha^2 + \cdots = 3$ in the limiting case.

The Sum to Infinity of a GP $= \frac{a}{1-r}$

In this case: $3 = \frac{1}{1-\alpha}$, which can be rearranged to find $\alpha = \frac{2}{3}$.

Question 6: B

Centre held fixed means that the equilibrium points on the wave (every half wavelength) must intersect with the centre as well as the ends. This occurs provided there is a whole number of waves in length L. The wavelength can therefore be any fraction of L.

Question 7: A

Conservation of energy: Kinetic energy lost = Work done (braking)

$\frac{1}{2}mu^2 = F_b d$ can be rearranged easily for F.

The sign of F will be negative due to the forward direction of u being positive.

Question 8: A

Put into completed square form.

$y = x^2 - 2x - 2$ becomes $y = (x - 1)^2 - 3$

y will be a minimum (-3) when the squared bracket is it's minimum (0).

Question 9: B

Use $m_t m_n = -1$ therefore at a right angle means $m = -\frac{1}{2}$.

Find y-coordinate of point of intersection: $y = 2(1) - 2 = 0$

Use $y - y_1 = m(x - x_1)$: $y - 0 = -\frac{1}{2}(x - 1)$

Question 10: E

Factorise (use factor theorem or trial and error): $(x - 1)^2(x + 1) = 0$

This gives the points of intersection with the x-axis as ± 1 (with $x = 1$ a point of contact rather than intersection as shown by the two roots in the same place).

Draw a sketch of a positive x^3 graph. It is above the x-axis continuously once it has passed through at $x = -1$.

Question 11:

Calculate the weight of the roof: $100(50)(100)(10) = 5,000,000N$

Calculate the area of the walls, remembering to prevent duplication of the corners.

$Area = 2(50)(0.1) + 2(100 - 0.2)(0.1) = 29.76m^2$

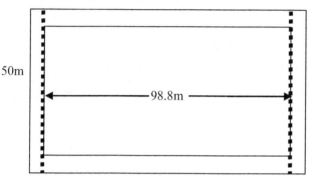

50m

98.8m

Convert area into mm^2 to match the units of stress: $2.976 \times 10^7 mm^2$

Calculate the stress using $\frac{Force}{Area} = \frac{5000000}{2.976 \times 10^7} = 0.17Nmm^{-2}$

Therefore all are suitable.

Question 12: C

Light bends towards the normal each time it enters a plate with greater refractive index. Therefore in the infinite limit it will be travelling along the normal.

Question 13:

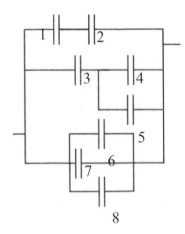

Combine the resistors a section at a time to calculate a total resistance. Use $C_t = C_a + C_b + \cdots$ for components in parallel, and $\frac{1}{C_t} = \frac{1}{C_a} + \frac{1}{C_b} + \cdots$ for components in series.

Combine 1 and 2 to get $\frac{C}{2}$.

Combine 4 and 5 to get $2R$. Then combine this with 3 for the total for that 'branch' as $\frac{2C}{3}$.

Combine 6, 7 and 8 to get $3C$.

Finally, combine the totals of all the 'branches' as $\frac{C}{2} + \frac{2C}{3} + 3C = 4\frac{1}{6}C$

Question 14:

Log law: $\log_b(a) = \frac{\log_c(a)}{\log_c(b)}$

First, convert the LHS to base 5 to get: $\log_x(25) = \frac{\log_5(25)}{\log_5(x)} = \log_5(x)$

Then multiply up: $\log_5(25) = (\log_5(x))^2$

Solve for x: $\log_5(x) = \pm\sqrt{2}$ so $x = 5^{\pm\sqrt{2}}$

Question 15:

a) Draw key points on graph, remembering times are given as the length of time at each speed, not the absolute value.

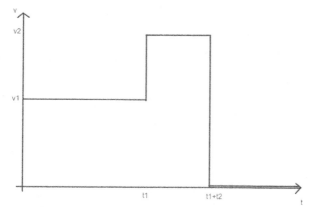

b) Average speed = total distance travelled / time moving. The total distance travelled is the area under the graph, giving:

$Av.\,speed = \frac{v_1\Delta t_1 + v_2\Delta t_2}{\Delta t_1 + \Delta t_2}$ and a value of: $\frac{(1)(2)+(2)(1)}{2+1} = \frac{4}{3}ms^{-1}$

c)

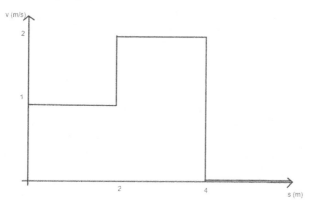

d) The distance weighted average is calculated similarly to the average speed as area under graph / total x-axis value:

$Distance\,weighted\,average\,speed = \frac{(1)(2)+(2)(2)}{2+2} = \frac{3}{2}ms^{-1}$

e) The average calculated in part b is the conventional definition.

f) Integrate

Question 16:

Area of circle $= \pi r^2$

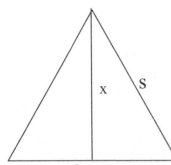

To find the area of the hexagon, split into 6 equilateral triangles with height x. $Area = \frac{1}{2}bh = \frac{1}{2}\frac{s}{2}x$

Find s in terms of x using Pythagoras: $x^2 + \left(\frac{s}{2}\right)^2 = s^2$ so $s = \frac{2x}{\sqrt{3}}$

$A_{triangle} = \frac{1}{4}\cdot\frac{2x^2}{\sqrt{3}} = \frac{x^2}{2\sqrt{3}}$ therefore

$A_{hexagon} = \frac{6x^2}{2\sqrt{3}} = \sqrt{3}x^2$

From the question, $\pi r^2 = \frac{\sqrt{3}x^2}{4}$, which can be rearranged to give $x = 2\sqrt{\frac{\pi}{\sqrt{3}}}r$

Question 17:

The integral: $\int_0^{0.1}(1+x)^9 dx = \left[\frac{1}{10}(1+x)^{10}\right]_0^{0.1}$

Taking the tenths outside, and letting x=0.1, gives; $\frac{1}{10}[(1+x)^{10}-1]$

This can be expanded to give: $(1+x)^{10} \approx 1 + 10x + \frac{10(9)}{2}x^2 + \cdots$

Therefore the integral can be 'evaluated' as:

$\frac{1}{10}[1 + 10x + 45x^2 + \cdots - 1] = x + 4.5x^2 + \cdots$

Expanding the brackets prior to integrating gives:

$(1+x)^9 \approx 1 + 9x + \frac{9(8)}{2}x^2 + \frac{9(8)(7)}{(3)(2)}x^3 + \cdots$

Which would integrate to $x + 4.5x^2 + \cdots$ and therefore only a first-order approximation is required to give an error less than 10%.

Question 18:

Draw a diagram – the diagram below shows a force balance on the left and dimensions given on the right (though both would be the same, as the system is symmetrical).

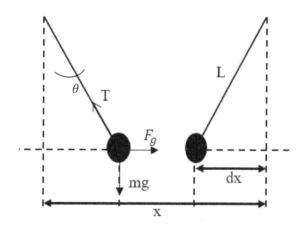

Consider only the forces on one of the balls.

Resolve vertically: $T\cos\theta = mg$

Resolve horizontally: $T\sin\theta = F_g$

Combine to eliminate T: $F_g = mg\tan\theta$

In order to find an expression for $\tan\theta$ first find the vertical height using Pythagoras: $h = \sqrt{L^2 - dx^2}$

Assumption: $L \gg dx$ so dx^2 will be so small that it can be ignored.

Therefore $h \approx L$ and $\tan\theta \approx \frac{dx}{L}$.

Substituting this into the expression for force gives $F_g = \frac{mgdx}{L}$.

Using Newton's Law of Gravitation $F_g = \frac{GMm}{r^2}$ which is written as

$F_g = \frac{Gm^2}{(x-2dx)^2}$ using the notation in the question.

Setting these equal to each other gives: $\frac{gdx}{L} = \frac{Gm}{(x-2dx)^2}$

Which can be rearranged to: $(x - 2dx)^2 dx = \frac{LmG}{g}$

Expanding the bracket gives: $x^2 dx - 4x dx^2 + 4 dx^3 = \frac{LmG}{g}$

Assumption: dx^3 is sufficiently small that it can be neglected.

This means the above simplifies to: $x^2 dx - 4x dx^2 = \frac{LmG}{g}$

Which can be rearranged to: $dx^2 - \frac{x}{4} dx = -\frac{LmG}{4xg}$

Complete the square: $\left(dx - \frac{x}{8} \right)^2 - \left(\frac{x}{8} \right)^2 = -\frac{LmG}{4xg}$

Rearrange and square root for the solution in the question.

b) For the wires to hang vertically the gravitational force and electric force must be equal and opposite: $\frac{Gm^2}{r^2} = k\frac{Q^2}{r^2}$

Therefore $Q^2 = \frac{Gm^2}{k}$ and $Q = m\sqrt{\frac{G}{k}}$

Question 19:

a) Resistance of A $= \frac{100^2}{100} = 10\Omega$, Resistance of B $= \frac{100^2}{20} = 50\Omega$

A greater proportion of the resistance will be dropped across bulb B (as it has a greater resistance), and therefore bulb B will be brightest.

Consider the bulbs as a potential divider such that voltage across each bulb is proportional to their contribution to the total resistance:

$V_A = \frac{10}{10+50} \cdot 100 = \frac{100}{6}$, $V_B = \frac{50}{50+10} \cdot 100 = \frac{500}{6}$

Finally, find the ratios of the powers (and brightness) with $P = \frac{V^2}{R}$

$\frac{\left(\frac{100}{6} \right)^2}{10} : \frac{\left(\frac{500}{6} \right)^2}{50}$ which simplifies to $\frac{1000}{6} : \frac{5000}{6}$ or $1:5$

(This shows that brightness is proportional to resistances – which you could, alternatively, have found algebraically).

b) If the lamps were in parallel, the lower resistance lamp would have a greater current flowing through and both bulbs would have the same voltage. Therefore the ratio of brightness would flip.

Question 20:

a) All the four planes have either 'all odd' or 'all even' (given the assumption that 0 is even). Therefore, it has an FCC structure.

b) Rearranging for 'a' gives: $a = d\sqrt{h^2 + k^2 + l^2}$

For each of the values given:

d (mm)	Plane	a (mm^2)
0.224	(111)	0.388
0.195	(200)	0.390
0.137	(220)	0.387
0.117	(311)	0.388

Therefore the best estimate is 0.39 to 2sf.

c) Volume before = $(Na)^3 = N^3 a^3$

Volume after = $\left(\dfrac{2}{3}Na\right) x^2$ where x is the length to be found.

Therefore: $x = \sqrt{\dfrac{N^3 a^3}{\frac{2}{3}Na}} = \sqrt{\dfrac{3}{2}}Na$

Question 21:

a) The minimum force would be an infinitesimal amount greater than the force required simply to keep the child's COM travelling in a circle, and is therefore $F > mr\omega^2$.

b) Work done $= \int_{r_0}^{0} mr\omega^2 dr$

c) The child's work is converted into kinetic energy, and as such the angular speed will increase as the child approaches the centre.

d) Because angular momentum is conserved, angular momentum 'J' before the child moves = angular momentum 'J' at any other point

J before $= \left(mr_0^2 + I_p\right)\omega_0$ (where ω_o denotes initial angular velocity.

J at a general point $= \left(mr^2 + I_p\right)\omega$

Setting these equal and rearranging gives: $\omega = \frac{(mr_0^2 + I_p)\omega_0}{mr^2 + I_p}$

e) Use the expression from part (d) in the integral from part (b) to give:

$$W.D = \int_{r_0}^{0} mr \left(\frac{\omega_0(mr_0^2 + I_p)}{mr^2 + I_p}\right)^2 dr$$

Taking constants outside the integral gives:

$$W.D = m\omega_0^2\left(mr_0^2 + I_p\right)^2 \int_{r_0}^{0} \frac{r}{(mr^2 + I_p)^2} dr$$

In order to get this to correspond to the 'hint' given in the question, the coefficient m in front of r^2 must be factorised out:

$$W.D = m\omega_0^2\left(mr_0^2 + I_p\right)^2 \int_{r_0}^{0} \frac{r}{(m^2)\left(r^2 + \frac{I_p}{m}\right)^2} dr$$

Taking the m^2 outside the integral as it is a constant allows the integration, with $r = 'x'$ and $\frac{I_p}{m} = 'a'$ in the hint for the integral of $\frac{x}{(a+x^2)^2}$. Therefore:

$$W.D = \frac{m\omega_0^2(mr_0^2 + I_p)^2}{m^2}\left[-\frac{1}{2\left(r^2 + \frac{I_p}{m}\right)}\right]_{r_0}^{0}$$

This gives the following expression:

$$W.D = \frac{\omega_0^2\left(mr_0^2 + I_p\right)^2}{m} \cdot -\frac{1}{2\left(\frac{I_p}{m}\right)} + \frac{\omega_0^2\left(mr_0^2 + I_p\right)^2}{m} \cdot \frac{1}{2\left(r_0^2 + \frac{I_p}{m}\right)}$$

This can be simplified significantly by cancelling terms, to give the final:

$$W.D = \frac{-mr_0^2\omega_0^2\left(mr_0^2 + I_p\right)}{2I_p}$$

Question 22:

It helps to visualise this question if a graph is drawn.

Firstly, rearrange the equation of the circle to allow it to be plotted.

This gives: $(x - 4)^2 + (y + 2)^2 = 16$ which is a circle, centre $(4, -2)$ radius 4.

The area to be found is therefore the one inside the circle but outside the triangle formed by the lines in the following diagram:

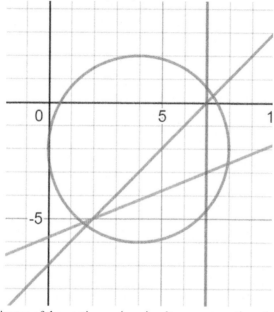

Find the coordinates of the vertices using simultaneous equations (and check against the graph):

$y = x - 7$ and $x = 7$ meet at $(7,0)$

$y = \frac{1}{5}(2x - 29)$ and $x = 7$ meet at $(7, -3)$

$y = x - 7$ and $y = \frac{1}{5}(2x - 29)$ can be set equal to give:

$x - 7 = \frac{1}{5}(2x - 29)$ and therefore $x - \frac{2}{5}x = 7 - \frac{29}{5}$

By combining like terms it can be found that they meet at $(2, -5)$.

Finding the area of the circle, using πr^2 gives 16π.

Find the area of the triangle using $\frac{bh}{2}$ (using base = 'purple distance' = 3, and height = 'horizontal distance' = 5). This gives $\frac{15}{2}$.

Therefore the area outside the triangle, but within the circle is $16\pi - \frac{15}{2}$.

Question 23:

$$f(x) = \frac{\sqrt{x^2 - 2}}{\ln(3x + 10)}$$

In order for this to be real: $x^2 - 2 \geq 0$ so $x \geq \sqrt{2}$ or $x \leq -\sqrt{2}$

In order for the denominator to be defined: $3x + 10 > 0$ so $x > -\frac{10}{3}$

In order for the function to be finite, the denominator must not be zero, as $\ln(1) = 0$ it must be that $3x + 10 \neq 1$ so $x \neq -3$.

Combining these gives: $-\frac{10}{3} < x \leq -\sqrt{2}$ or $x \geq \sqrt{2}$ and $x \neq -3$

END OF PAPER

Afterword

Remember that the route to a high score is your approach and practice. Don't fall into the trap that *"you can't prepare for the PAT"*– this couldn't be further from the truth. With knowledge of the test, time-saving techniques and plenty of practice you can dramatically boost your score.

Work hard, never give up and do yourself justice.

Good luck!

Acknowledgements

Thanks must go to *Samuel* for his tremendous help in putting this set of answers together.

Rohan

About UniAdmissions

UniAdmissions is an educational consultancy that specialises in supporting **applications to Medical School and to Oxbridge**.

Every year, we work with hundreds of applicants and schools across the UK. From free resources to our *Ultimate Guide Books* and from intensive courses to bespoke individual tuition – with a team of **300 Expert Tutors** and a proven track record, it's easy to see why UniAdmissions is the **UK's number one admissions company**.

To find out more about our support like intensive **courses** and **tuition**, check out **www.uniadmissions.co.uk/PAT**

YOUR FREE BOOK

Thanks for purchasing this Ultimate Guide Book. Readers like you have the power to make or break a book – hopefully you found this one useful and informative. *UniAdmissions* would love to hear about your experiences with this book.

As thanks for your time we'll send you another ebook from our Ultimate Guide series absolutely <u>FREE</u>!

How to Redeem Your Free Ebook

1) Find the book you have either on your Amazon purchase history or your email receipt to help find the book on Amazon.

2) On the product page at the Customer Reviews area, click 'Write a customer review'. Write your review and post it! Copy the review page or take a screen shot of the review you have left.

3) Head over to www.uniadmissions.co.uk/free-book and select your chosen free ebook! You can choose from:

- ➤ The Ultimate PAT Guide
- ➤ The Ultimate Oxbridge Interview Guide
- ➤ The Ultimate UCAS Personal Statement Guide
- ➤ PAT Past Paper Solutions

Your ebook will then be emailed to you – it's as simple as that! Alternatively, you can buy all the above titles at **www.uniadmisions.co.uk/our-books**

OXBRIDGE INTERVIEW COURSE

If you've got an upcoming interview for Oxford – this is the perfect course for you. You get individual attention throughout the day and are taught by specialist Oxbridge graduates on how to approach these tricky interviews.

Full Day intensive Course
Guaranteed Small Groups
4 Hours of Small group teaching
4 x 30 minute individual Mock Interviews
Full written feedback so you can see how to improve
Ongoing Tutor Support until your interview – never be alone again

Timetable:

1000 - 1015: Registration
1015 - 1030: Talk: Key to interview Success
1030 - 1130: Tutorial: Dealing with Unknown Material
1145 - 1245: 2 x Individual Mock Interviews
1245 - 1330: Lunch
1330 - 1430: Subject Specific Tutorial
1445 - 1545: 2 x Individual Mock Interviews
1600 - 1645: Subject Specific Tutorial
1645 - 1730: Debrief and Finish

The course is normally £395 but you can get £35 off by using the code *"BRK35"* at checkout.

www.uniadmissions.co.uk/oxbridge-interview-course

£35 VOUCHER:

BRK35

Printed in Great Britain
by Amazon

62076172R00088